AF306381

Monographs in

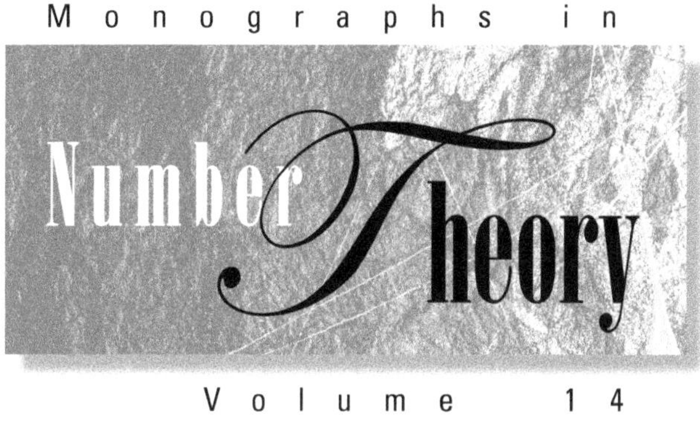

Volume 1 4

Jacobsthal Sums

Monographs in

Volume 14

Jacobsthal Sums

Bogdan Nica

Indiana University Indianapolis, USA

World Scientific

NEW JERSEY • LONDON • SINGAPORE • BEIJING • SHANGHAI • TAIPEI • CHENNAI

Published by

World Scientific Publishing Co. Pte. Ltd.
5 Toh Tuck Link, Singapore 596224
USA office: 27 Warren Street, Suite 401-402, Hackensack, NJ 07601
UK office: 57 Shelton Street, Covent Garden, London WC2H 9HE

Library of Congress Control Number: 2025013059

British Library Cataloguing-in-Publication Data
A catalogue record for this book is available from the British Library.

Monographs in Number Theory — Vol. 14
JACOBSTHAL SUMS

ISBN 9789819813162 (hardcover)
ISBN 9789819813179 (ebook for institutions)
ISBN 9789819813186 (ebook for individuals)

For any available supplementary material, please visit
https://www.worldscientific.com/worldscibooks/10.1142/14313#t=suppl

Desk Editors: Eshak Nabi Akbar Ali/Rok Ting Tan

Typeset by Stallion Press
Email: enquiries@stallionpress.com

Preface

What is a Jacobsthal sum? Let p be an odd prime. The two sums

$$\sum_{x=0}^{p-1}\left(\frac{x^3+1}{p}\right), \qquad \sum_{x=0}^{p-1}\left(\frac{x^3+x}{p}\right) \tag{1}$$

epitomize the Jacobsthal sums considered in this text. The above sums involve the Legendre symbol, which records the quadratic nature modulo p of a given integer as follows: when n is relatively prime to p, we set

$$\left(\frac{n}{p}\right) = \begin{cases} 1 & \text{if } n \text{ is a quadratic residue mod } p, \\ -1 & \text{if } n \text{ is a quadratic non-residue mod } p. \end{cases}$$

When n is a multiple of p, we set $(n/p) = 0$.

Why are Jacobsthal sums interesting? Sums of Legendre symbols with polynomial arguments naturally emerge when counting the number of solutions to equations over the finite field $\mathbb{Z}/p\mathbb{Z}$, which we systematically denote by $\mathbb{F}_p$ in what follows. The most direct manifestation of this idea concerns the equation $y^2 = f(x)$. For each $x \in \mathbb{F}_p$, there are $1 + (f(x)/p)$ solutions $y \in \mathbb{F}_p$. Summing over $x \in \mathbb{F}_p$, we obtain the following formula for the number of solutions:

$$\# \left\{ (x,y) \in \mathbb{F}_p \times \mathbb{F}_p : y^2 = f(x) \right\} = p + \sum_{x \in \mathbb{F}_p} \left(\frac{f(x)}{p}\right).$$

Evaluating the latter sum is thereby revealed to be the heart of the matter. In particular, the Jacobsthal sums in (1) essentially count the number solutions to the equation $y^2 = x^3 + 1$, respectively, the equation $y^2 = x^3 + x$, over $\mathbb{F}_p$. These are typical examples of so-called elliptic equations.

The Jacobsthal sums (1) also appear in other solution-counting problems over $\mathbb{F}_p$ such as the quartic Fermat equation $x^4 + y^4 = 1$; the "last entry" equation $(1 + x^2)(1 + y^2) = 2$; the Lehmers' equation $x^{-1} + y^{-1} + z^{-1} = x + y + z$. In these examples, the appearance of the sums (1) is less evident; on occasion, it can be truly concealed–our discussion of Lehmers' equation being a pointed illustration.

There goes the first reason why Jacobsthal sums are interesting: they are very useful for counting the number of solutions to equations over the finite field $\mathbb{F}_p$. The second reason why Jacobsthal sums are interesting has to do with representations of p as a sum of two integral squares.

An early jewel of number theory is the following result of Pierre Fermat: if a prime p satisfies $p \equiv 1 \bmod 4$, then $p = A^2 + B^2$ for some integers A and B. Facts as old as this one have several proofs–typically non-constructive, just as the statement would suggest. But Jacobsthal sums offer an *explicit* proof of Fermat's theorem. Namely,

$$A = \frac{1}{2} \sum_{x=0}^{p-1} \left(\frac{x^3 + x}{p} \right), \qquad B = \frac{1}{2} \sum_{x=0}^{p-1} \left(\frac{x^3 + bx}{p} \right),$$

where b is a quadratic non-residue modulo p, turn out to be integers satisfying the relation $A^2 + B^2 = p$. Two other representations by sums of squares, also put forth by Fermat, can be made explicit in a similar way by means of Jacobsthal sums. These are the following: if $p \equiv 1 \bmod 6$, then $p = A^2 + 3B^2$ for some integers A and B; if $p \equiv 1 \bmod 8$, then $p = A^2 + 2B^2$ for some integers A and B.

This text. The story of Jacobsthal sums, as told in this short monograph, has several subplots: (i) the general properties of Jacobsthal sums, (ii) sum-of-squares representations for p and evaluations of Jacobsthal sums, (iii) mod p evaluations of some binomial coefficients, (iv) the use of Jacobsthal sums in counting solutions to equations over $\mathbb{F}_p$, and (v) the interplay with the family of ϱ-sums.

The text starts with two preliminary chapters (Chapters 1 and 2). Items (i), (ii), and (iii) are covered in Chapter 3. Chapter 4 illustrates item (iv) with an abundance of examples. In Chapter 5, we introduce ϱ-sums, we discuss some of their properties, and we illustrate their use–both for understanding other sums, and for solution counting; along the way, we establish several points of contact with Jacobsthal sums.

The only other systematic treatment of Jacobsthal sums can be found in the outstanding reference *Gauss and Jacobi Sums* (John Wiley & Sons 1998) by Bruce Berndt, Ronald Evans, and Kenneth Williams. There is some overlap between Chapter 6 of that monograph and Chapter 3 of this text. But Chapters 4 and 5 set this text apart. The great emphasis we place on using Jacobsthal sums, and related quadratic character sums, for solution counting is a distinctive feature. Another novel viewpoint is the cohesive discussion of ϱ-sums.

We have strived to get as far as possible while staying self-contained and elementary. Very intentionally, we have confined ourselves: in this text we focus on quadratic characters sums over prime fields, and we avoid overreaching into general finite fields, or into other types of character sums such as Jacobi sums.

Each chapter ends with Exercises and Notes. We have devoted a great deal of attention to selecting and creating relevant exercises; our judgement is that many of them are non-trivial. The final part of this text collects solutions to all exercises. The notes at the end of each chapter discuss context and indicate references.

Acknowledgments. I thank Pete Clark, Darij Grinberg, Patrick Morton, Kenneth Williams, and the anonymous referees for feedback on preliminary versions of this text. And I will be grateful to you, dear reader, for sending me your feedback at bnica@iu.edu.

About the Author

Bogdan Nica earned his degrees at McGill University (BSc, MSc) and Vanderbilt University (PhD). After temporary positions at the University of Victoria, the University of Göttingen, and McGill University, he joined the faculty at Indiana University Indianapolis in 2020.

This is his second book. His first book, *A Brief Introduction to Spectral Graph Theory*, has been published by the European Mathematical Society in 2018; a Japanese translation has appeared in 2024.

Author's address

Bogdan Nica

Department of Mathematical Sciences
Indiana University Indianapolis, USA

MSC2020 Classification

Primary:

11L10 Jacobsthal and Brewer sums; other complete character sums

Secondary:

11G20 Curves over finite and local fields
11G25 Varieties over finite and local fields
11D45 Counting solutions of Diophantine equations
11D79 Congruences in many variables
11-02 Research exposition pertaining to number theory

Contents

Chapter 1

Preliminaries

Throughout this text, p is an odd prime.

Powers

Consider $\mathbb{F}_p = \mathbb{Z}/p\mathbb{Z}$, a finite field with p elements. The set of non-zero elements in $\mathbb{F}_p$, denoted $\mathbb{F}_p^*$, is a group under multiplication. Its order is $p-1$; consequently, $x^{p-1} = 1$ for all $x \in \mathbb{F}_p^*$.

The multiplicative group $\mathbb{F}_p^*$ is, in fact, cyclic. A generator of $\mathbb{F}_p^*$ is also known as a primitive root mod p. To paraphrase a memorable idea of Hermann Weyl, the introduction of a generator in an argument concerning $\mathbb{F}_p^*$ is a small act of violence–but what a convenient and effective act it is. This is especially conspicuous when considering powers in $\mathbb{F}_p^*$.

Theorem 1.1. *For every integer n, we have*

$$\sum_{x \in \mathbb{F}_p^*} x^n = \begin{cases} -1 & \text{if } n \text{ is a multiple of } p-1, \\ 0 & \text{otherwise.} \end{cases}$$

Proof. Let S_n denote the above sum of powers. If $p-1$ divides n then $x^n = 1$ for each $x \in \mathbb{F}_p^*$, and so $S_n = p-1 = -1$ in $\mathbb{F}_p$. If $p-1$ does not divide n, then let g be a generator of $\mathbb{F}_p^*$ and observe that

$$g^n S_n = \sum_{x \in \mathbb{F}_p^*} (gx)^n = \sum_{x \in \mathbb{F}_p^*} x^n = S_n.$$

As $g^n \neq 1$, it follows that $S_n = 0$. $\qquad\square$

Corollary 1.2. *Let*

$$f(x) = \sum_{k=0}^{n} c_k x^k$$

be a polynomial with coefficients in $\mathbb{F}_p$. Then

$$\sum_{x \in \mathbb{F}_p} f(x) = - \sum_{\substack{k=1 \\ p-1 \mid k}}^{n} c_k.$$

Next, we describe which elements in $\mathbb{F}_p^*$ are nth powers, and in how many ways.

Theorem 1.3. *Let $a \in \mathbb{F}_p^*$. Then the equation $x^n = a$ is solvable in $\mathbb{F}_p^*$ if and only if $a^{(p-1)/d} = 1$, where $d = \gcd(n, p-1)$. There are $(p-1)/d$ such values of $a \in \mathbb{F}_p^*$, and for each one the number of solutions to the equation $x^n = a$ is d.*

Proof.　Assume that the equation $x^n = a$ is solvable in $\mathbb{F}_p^*$. Then, thinking of $x \in \mathbb{F}_p^*$ as a particular solution, we have $a^{(p-1)/d} = x^{(p-1)n/d} = 1$.

Conversely, assume that $a^{(p-1)/d} = 1$. We show that $x^n = a$ has d solutions; in particular, the equation is solvable. Using $n = (p-1)/d$, it also follows that there are $(p-1)/d$ values of $a \in \mathbb{F}_p^*$ satisfying $a^{(p-1)/d} = 1$.

Let $g \in \mathbb{F}_p^*$ be a generator, so $a = g^r$, where $0 \le r \le p-1$. From $a^{(p-1)/d} = g^{r(p-1)/d} = 1$, we deduce that d divides r. Put $r = dr'$. Also, let $n = dn'$. Now, $x = g^k \in \mathbb{F}_p^*$ satisfies $x^n = a$ if and only if $g^{kn} = g^r$, if and only if $kn \equiv r \bmod p-1$, if and only if $kn' \equiv r' \bmod (p-1)/d$. As n' is relatively prime to $(p-1)/d$, the latter condition specifies $k \bmod (p-1)/d$, so there are exactly d possible values of k in the range $0, \ldots, p-1$.　□

Corollary 1.4. *The power function $x \mapsto x^n$ is a permutation on $\mathbb{F}_p$ if and only if n is relatively prime to $p-1$.*

Corollary 1.5. *Assume $p \equiv 1 \bmod n$. Then the equation $x^n = -1$ is solvable in $\mathbb{F}_p$ if and only if $p \equiv 1 \bmod 2n$.*

The quadratic character

An element $a \in \mathbb{F}_p^*$ is a *square* if the equation $x^2 = a$ is solvable in $\mathbb{F}_p^*$, and a *non-square* otherwise. Integers that are not divisible by p are called *quadratic residues*, respectively, *quadratic non-residues*, modulo p according to whether they are squares or non-squares modulo p.

Let us spell out what Theorem 1.3 says for $n = 2$.

Corollary 1.6. *Let $a \in \mathbb{F}_p^*$. Then a is a square if and only if $a^{(p-1)/2} = 1$. There are $(p-1)/2$ such values of $a \in \mathbb{F}_p^*$, and for each one of them the equation $x^2 = a$ has two solutions.*

Thus, in $\mathbb{F}_p^*$, there are $(p-1)/2$ squares and $(p-1)/2$ non-squares.

The *quadratic character* σ on $\mathbb{F}_p$ is the integer-valued map defined as follows: for $a \in \mathbb{F}_p$, we set

$$\sigma(a) = \begin{cases} 1 & \text{if } a \text{ is a square in } \mathbb{F}_p^*, \\ -1 & \text{if } a \text{ is a non-square in } \mathbb{F}_p^*, \\ 0 & \text{if } a = 0. \end{cases} \tag{2}$$

The quadratic character is a close relative of the Legendre symbol $(\cdot/p)$. The notation, σ, echoes its role–that of recording the "squaredness"–, as well as its nature–it is a signature, being essentially ± 1-valued.

Example 1.7. We have

$$\sigma(-1) = \begin{cases} 1 & \text{if } p \equiv 1 \bmod 4, \\ -1 & \text{if } p \equiv 3 \bmod 4. \end{cases} \tag{3}$$

Indeed, the equation $x^2 = -1$ is solvable if and only if $p \equiv 1 \bmod 4$. This is the case $n = 2$ of Corollary 1.5.

Example 1.8. Let $p \neq 3$. We have

$$\sigma(-3) = \begin{cases} 1 & \text{if } p \equiv 1 \bmod 3, \\ -1 & \text{if } p \equiv 2 \bmod 3. \end{cases} \tag{4}$$

If $p \equiv 1 \bmod 3$ then the equation $x^3 = 1$ has three solutions in $\mathbb{F}_p$, but it has only one solution, namely $x = 1$, if $p \equiv 2 \bmod 3$. Thus $x^2 + x + 1 = 0$ is solvable in $\mathbb{F}_p$ if and only if $p \equiv 1 \bmod 3$. Now, $x^2 + x + 1 = 0$ is equivalent to $4x^2 + 4x + 4 = 0$, that is, $(2x + 1)^2 = -3$. Visibly, the latter equation is solvable if and only if -3 is a square in $\mathbb{F}_p^*$.

The following theorem collects some basic properties of the quadratic character.

Theorem 1.9. *The following hold:*

(i) *(Euler's identity) $\sigma(a) \equiv a^{(p-1)/2} \bmod p$ for each $a \in \mathbb{F}_p$;*
(ii) *(Multiplicativity) $\sigma(ab) = \sigma(a)\sigma(b)$ for all $a, b \in \mathbb{F}_p$;*

(iii) *the equation $x^2 = a$ has $1 + \sigma(a)$ solutions in $\mathbb{F}_p$; more generally, the quadratic equation $x^2 + bx + c = 0$ has $1 + \sigma(\Delta)$ solutions in $\mathbb{F}_p$, where $\Delta = b^2 - 4c$.*

Proof. (i) For $a \in \mathbb{F}_p^*$, we have $(a^{(p-1)/2})^2 = a^{p-1} = 1$ so $a^{(p-1)/2} = \pm 1$. As we have already noted, $a^{(p-1)/2} = 1$ if and only if a is a square. So, if $\sigma(a) = 1$ then $a^{(p-1)/2} = 1$; if $\sigma(a) = -1$ then $a^{(p-1)/2} = -1$; whereas if $a = 0$, we have $\sigma(a) = 0$ and $a^{(p-1)/2} = 0$. In all cases, $\sigma(a)$ agrees with $a^{(p-1)/2}$. There is a distinction to be made, however: $\sigma(a) = 0, \pm 1$ in $\mathbb{Z}$ while $a^{(p-1)/2} = 0, \pm 1$ in $\mathbb{F}_p = \mathbb{Z}/p\mathbb{Z}$. We conclude that $\sigma(a)$ modulo p equals $a^{(p-1)/2}$; slightly abusively, we write this as $\sigma(a) \equiv a^{(p-1)/2} \bmod p$.

(ii) Using Euler's identity, we have

$$\sigma(ab) \equiv (ab)^{(p-1)/2} = a^{(p-1)/2}\, b^{(p-1)/2} \equiv \sigma(a)\sigma(b) \quad \bmod p.$$

As the possible values of $\sigma(ab)$ and $\sigma(a)\sigma(b)$ are 0, ± 1, from $\sigma(ab) \equiv \sigma(a)\sigma(b) \bmod p$ we infer that $\sigma(ab) = \sigma(a)\sigma(b)$.

(iii) If a is a square in $\mathbb{F}_p^*$, then the equation $x^2 = a$ has $2 = 1 + \sigma(a)$ solutions. If a is a non-square in $\mathbb{F}_p^*$, then the equation $x^2 = a$ has $0 = 1 + \sigma(a)$ solutions. Finally, if $a = 0$ then the number of solutions to $x^2 = a$ is $1 = 1 + \sigma(a)$. In all cases, the equation $x^2 = a$ has $1 + \sigma(a)$ solutions.

The equation $x^2 + bx + c = 0$ is equivalent, after the change of variable $x := x - b/2$, to $x^2 = \Delta$; this equation has $1 + \sigma(\Delta)$ solutions in $\mathbb{F}_p$. $\square$

The multiplicativity of the quadratic character amounts, essentially, to the fact that σ is a group homomorphism from the multiplicative group $\mathbb{F}_p^*$ to the multiplicative group $\{\pm 1\}$. Clearly, σ is onto as well. It is not hard to see that σ is, in fact, the unique group homomorphism from $\mathbb{F}_p^*$ onto $\{\pm 1\}$.

We now turn to some simple but useful observations concerning sums that involve the quadratic character σ.

Theorem 1.10. *We have*

$$\sum_{a \in \mathbb{F}_p} \sigma(a) = 0. \tag{5}$$

Proof. We give three arguments for the vanishing of the sum

$$S = \sum_{a \in \mathbb{F}_p} \sigma(a).$$

First argument. Recall that there is a fair share of squares and non-squares in $\mathbb{F}_p^*$–namely, $(p-1)/2$ of each kind. Therefore,

$$S = \sigma(0) + \#\{a \in \mathbb{F}_p^* : \sigma(a) = 1\} - \#\{a \in \mathbb{F}_p^* : \sigma(a) = -1\} = 0.$$

Second argument. Let $c \in \mathbb{F}_p^*$. The change of variable $a := ca$ gives

$$S = \sum_{a \in \mathbb{F}_p} \sigma(ca) = \sigma(c) \sum_{a \in \mathbb{F}_p} \sigma(a) = \sigma(c)S.$$

Choosing c to be a non-square, we deduce that $S = 0$.

Third argument. By Euler's identity, we have

$$S \equiv \sum_{a \in \mathbb{F}_p} a^{(p-1)/2} \mod p.$$

The latter sum vanishes in $\mathbb{F}_p$, thanks to Corollary 1.2, so $S \equiv 0 \mod p$. However, the terms in S are ± 1 except for $\sigma(0) = 0$, implying that S is an integer in the interval $[-(p-1),(p-1)]$. We conclude that $S = 0$. $\square$

Theorem 1.11. *We have*

$$\sum_{\substack{a,b \in \mathbb{F}_p \\ a+b=c}} \sigma(a)\sigma(b) = \begin{cases} -\sigma(-1) & \text{if } c \neq 0, \\ \sigma(-1)(p-1) & \text{if } c = 0. \end{cases} \tag{6}$$

Proof. For each $c \in \mathbb{F}_p$, we have

$$\sum_{\substack{a,b \in \mathbb{F}_p \\ a+b=c}} \sigma(a)\sigma(b) = \sum_{a \in \mathbb{F}_p} \sigma\big(a(c-a)\big) = \sum_{a \in \mathbb{F}_p^*} \sigma\big(a(c-a)\big).$$

When $c = 0$, each term equals $\sigma(-1)$, whence the sum equals $(p-1)\sigma(-1)$. When $c \neq 0$, we make the change of variable $a := ca^{-1}$ in the above sum. As $\sigma\big(ca^{-1}(c - ca^{-1})\big) = \sigma\big((ca^{-1})^2(a-1)\big) = \sigma(a-1)$, the above sum becomes

$$\sum_{a \in \mathbb{F}_p^*} \sigma(a-1) = -\sigma(-1),$$

thanks to (5). $\square$

Example 1.12. Assume $p \equiv 1 \mod 4$. The *Paley graph* of order p has the elements of $\mathbb{F}_p$ as vertices. Edges join two distinct vertices x and y whenever $x - y$ is a square in $\mathbb{F}_p$; two vertices which are joined by an edge are said to be neighbors. We note that the edge rule is symmetric in x and y, since -1 is a square in $\mathbb{F}_p$. We also note that the Paley graph of order p is regular of degree $(p-1)/2$, that is to say, each vertex has $(p-1)/2$ neighbors. This reflects the fact that there are $(p-1)/2$ squares in $\mathbb{F}_p^*$.

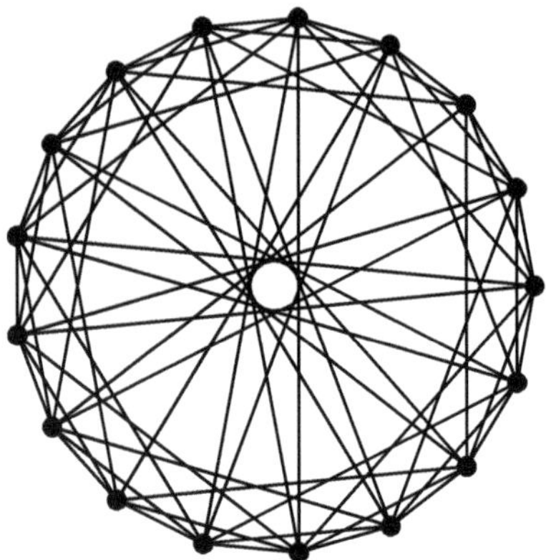

The Paley graph of order 17.

Interestingly, the Paley graph of order p is strongly regular, in the following sense: the number of common neighbors that two distinct vertices have, only depends on whether the two vertices are neighbors or not.

Indeed, let $x, y \in \mathbb{F}_p$ be distinct vertices in the Paley graph of order p. A vertex $z \in \mathbb{F}_p$ is a common neighbor of x and y precisely when $\sigma(x-z) = \sigma(z-y) = 1$. So, the expression $\big(1+\sigma(x-z)\big)\big(1+\sigma(z-y)\big)$ equals 4 when z is a common neighbor of x and y; it equals $1 + \sigma(x-y)$ if $z = x$ or $z = y$; otherwise, it vanishes. We may thus count by a means of a sum which involves the quadratic character:

$$\sum_{z \in \mathbb{F}_p} \big(1+\sigma(x-z)\big)\big(1+\sigma(z-y)\big) = 2\big(1+\sigma(x-y)\big) + 4N_{xy},$$

where N_{xy} denotes the number of common neighbors of x and y. The left-hand sum breaks into four sums

$$\sum_{z \in \mathbb{F}_p} 1, \qquad \sum_{z \in \mathbb{F}_p} \sigma(x-z), \qquad \sum_{z \in \mathbb{F}_p} \sigma(z-y), \qquad \sum_{z \in \mathbb{F}_p} \sigma(x-z)\sigma(z-y)$$

which evaluate as follows: the first one equals p; the next two vanish, being (5) in disguise; the last one equals $-\sigma(-1) = -1$, thanks to Theorem 1.11.

Therefore,

$$N_{xy} = \frac{p - 3 - 2\sigma(x - y)}{4} = \begin{cases} (p-5)/4 & \text{if } x \text{ and } y \text{ are neighbors,} \\ (p-1)/4 & \text{if } x \text{ and } y \text{ are not neighbors.} \end{cases}$$

Convolutions and solution-counting maps

Let us consider, very broadly, numerical maps on $\mathbb{F}_p$–that is to say, integer-valued functions defined on $\mathbb{F}_p$. There is an obvious operation of addition for numerical maps on $\mathbb{F}_p$; it is defined by the rule $(\alpha + \beta)(c) = \alpha(c) + \beta(c)$ for each $c \in \mathbb{F}_p$. We could consider an operation of multiplication by an analogous, pointwise rule. A more interesting multiplication is the *convolution* of two numerical maps, defined by

$$(\alpha * \beta)(c) = \sum_{\substack{a,b\in\mathbb{F}_p \\ a+b=c}} \alpha(a)\beta(b). \tag{7}$$

It is easy to see that the operation of convolution has the following properties: (i) it is commutative, $\alpha * \beta = \beta * \alpha$; (ii) it is associative, $(\alpha * \beta) * \gamma = \alpha * (\beta * \gamma)$; (iii) it distributes over addition, $\alpha * (\beta + \gamma) = \alpha * \beta + \alpha * \gamma$; (iv) the map δ_0 given by

$$\delta_0(c) = \begin{cases} 1 & \text{if } c = 0 \\ 0 & \text{if } c \neq 0 \end{cases}$$

is a multiplicative identity: $\alpha * \delta_0 = \delta_0 * \alpha = \alpha$. In short, numerical maps on $\mathbb{F}_p$ form a commutative ring with identity under addition and convolution.

Let $\mathbb{1}$ denote the constant 1 map on $\mathbb{F}_p$. We note that, for any numerical map α on $\mathbb{F}_p$, the convolution $\alpha * \mathbb{1}$ is constant, namely,

$$\alpha * \mathbb{1} = \sum_{a\in\mathbb{F}_p} \alpha(a). \tag{8}$$

Although δ_0 plays a distinguished role for convolution, we will find it more convenient to work with the map $\kappa = p\delta_0 - \mathbb{1}$. Explicitly,

$$\kappa(c) = \begin{cases} p - 1 & \text{if } c = 0, \\ -1 & \text{if } c \neq 0. \end{cases} \tag{9}$$

Lemma 1.13. *We have*

$$1 * 1 = p1, \quad 1 * \kappa = 0, \quad \kappa * \kappa = p\kappa$$

and

$$1 * \sigma = 0, \quad \kappa * \sigma = p\sigma, \quad \sigma * \sigma = \sigma(-1)\kappa.$$

Proof. The formulas for $1 * 1$, $1 * \kappa$, and $1 * \sigma$ can be deduced directly from (8). The formulas for convolution against κ, namely $\kappa * \kappa$ and $\kappa * \sigma$, are then obtained using the following:

$$\kappa * \alpha = (p\delta_0 - 1) * \alpha = p(\delta_0 * \alpha) - (1 * \alpha) = p\alpha - (1 * \alpha).$$

The formula for $\sigma * \sigma$ is the convolutional rephrasing of Theorem 1.11. $\square$

Convolution calculus has a direct bearing on solution counting. Let $f : \mathbb{F}_p^n \to \mathbb{F}_p$ be a function in n variables, $f = f(x_1, \ldots, x_n)$. The *solution-counting map* for f is the numerical map N_f, defined by

$$N_f(c) = \#\{(x_1, \ldots, x_n) \in \mathbb{F}_p^n : f(x_1, \ldots, x_n) = c\} \tag{10}$$

for each $c \in \mathbb{F}_p$. Concisely, $N_f(c) = \#f^{-1}(c)$. Since the preimages $\{f^{-1}(c) : c \in \mathbb{F}_p\}$ partition $\mathbb{F}_p^n$, we have

$$\sum_{c \in \mathbb{F}_p} N_f(c) = p^n. \tag{11}$$

In statistical language this means that, for a function in n variables over $\mathbb{F}_p$, the expected value of its solution-counting map is p^{n-1}.

Lemma 1.14 (Separation principle). *Given two functions $f : \mathbb{F}_p^n \to \mathbb{F}_p$ and $g : \mathbb{F}_p^m \to \mathbb{F}_p$, let $f + g : \mathbb{F}_p^{n+m} = \mathbb{F}_p^n \times \mathbb{F}_p^m \to \mathbb{F}_p$ be defined as $(f+g)(x, y) = f(x) + g(y)$. Then the solution-counting map for the function $f + g$ is given by*

$$N_{f+g} = N_f * N_g.$$

Proof. Let $c \in \mathbb{F}_p$. The solution set $\{(x, y) \in \mathbb{F}_p^n \times \mathbb{F}_p^m : f(x) + g(y) = c\}$ is the disjoint union, over all additive decompositions $c = a + b$, of product sets of the form $\{x \in \mathbb{F}_p^n : f(x) = a\} \times \{y \in \mathbb{F}_p^m : g(y) = b\}$. Taking cardinalities, we find

$$N_{f+g}(c) = \sum_{a+b=c} N_f(a)N_g(b),$$

that is $N_{f+g}(c) = (N_f * N_g)(c)$. $\square$

Example 1.15. A diagonal equation is an equation of the form

$$a_1 x^{d_1} + a_2 x^{d_2} + \cdots + a_n x^{d_n} = c,$$

where $a_1, \ldots, a_n \in \mathbb{F}_p^*$, $c \in \mathbb{F}_p$, and $d_1, \ldots, d_n$ are positive integers.

Put $e_i = \gcd(d_i, p-1)$ for each i. We can infer from Theorem 1.3 that the power functions x^{d_i} and x^{e_i} have the same solution-counting maps. It follows, by the separation principle, that the functions $a_1 x^{d_1} + \cdots + a_n x^{d_n}$ and $a_1 x^{e_1} + \cdots + a_n x^{e_n}$ have the same solution-counting maps as well.

The upshot is that, in what concerns solution counts for diagonal equations over $\mathbb{F}_p$, we may assume without loss of generality that the exponents divide $p-1$.

The quadratic equation $ax^2 + by^2 = c$

Theorem 1.16. *Let $a, b \in \mathbb{F}_p^*$, and $c \in \mathbb{F}_p$. Then the equation*

$$ax^2 + by^2 = c \tag{12}$$

has $p - \sigma(-ab)$ solutions for $c \neq 0$, and $p + \sigma(-ab)(p-1)$ solutions for $c = 0$.

The theorem says that the solution-counting map for $ax^2 + by^2$ is nearly constant. As for the proof, it is a nice illustration of the separation principle.

Proof. The solution-counting map $N_{ax^2+by^2}$ can be computed as a convolution, $N_{ax^2+by^2} = N_{ax^2} * N_{by^2}$. Now,

$$N_{ax^2}(c) = \#\{x \in \mathbb{F}_p : ax^2 = c\} = 1 + \sigma(ac);$$

in functional form, $N_{ax^2} = \mathbb{1} + \sigma(a)\sigma$. Similarly, $N_{by^2} = \mathbb{1} + \sigma(b)\sigma$. Therefore,

$$N_{ax^2+by^2} = \big(\mathbb{1} + \sigma(a)\sigma\big) * \big(\mathbb{1} + \sigma(b)\sigma\big) = p\mathbb{1} + \sigma(-ab)\kappa. \tag{13}$$

Explicitly, $N_{ax^2+by^2}(c) = p - \sigma(-ab)$ for $c \neq 0$, respectively, $N_{ax^2+by^2}(0) = p + \sigma(-ab)(p-1)$. $\qquad\square$

We consider two applications of Theorem 1.16.

Example 1.17. The equation

$$x^2 + y^2 = 1$$

has $N = p - \sigma(-1)$ solutions. But a closer look at the solution set will lead us to an interesting outcome.

The solution set has two sign symmetries, $(x, y) \mapsto (-x, y)$ and $(x, y) \mapsto (x, -y)$. The geometric picture is that of a mod p "circle", symmetric with respect to the x- and y-axes. Thus, the solution set can be partitioned into 4-tuples

$$\{(x, y), (-x, y), (x, -y), (-x, -y)\}$$

except for the four intercepts $(\pm 1, 0)$, $(0, \pm 1)$. The counting upshot of this viewpoint is that $N \equiv 0 \bmod 4$. Therefore, $\sigma(-1) \equiv p \bmod 4$, whence

$$\sigma(-1) = \begin{cases} 1 & \text{if } p \equiv 1 \bmod 4, \\ -1 & \text{if } p \equiv 3 \bmod 4; \end{cases}$$

in compact form:

$$\sigma(-1) = (-1)^{(p-1)/2}. \tag{14}$$

The evaluation of $\sigma(-1)$ is an old fact, Example 1.7, in a new light. But let us dig further. If we consider yet another symmetry, the flip $(x, y) \mapsto (y, x)$, then the solution set can be partitioned into 8-tuples of the form

$$\{(x, y), (-x, y), (x, -y), (-x, -y), (y, x), (-y, x), (y, -x), (-y, -x)\}.$$

Well, nearly: such a generic 8-tuple may degenerate into a 4-tuple. One case is the axial 4-tuple $\{(\pm 1, 0), (0, \pm 1)\}$. The other potential exception arises from solutions satisfying $y = \pm x$; these solutions would yield the diagonal 4-tuple $\{(x, x), (x, -x), (-x, x), (-x, -x)\}$, where $2x^2 = 1$. The latter 4-tuple occurs if and only if 2 is a square in $\mathbb{F}_p$, so we may account for its size as $2 + 2\sigma(2)$. The refined counting upshot is now that

$$N \equiv 4 + (2 + 2\sigma(2)) \equiv 2\sigma(2) - 2 \bmod 8.$$

On the other hand, the formula $N = p - \sigma(-1)$ together with the evaluation of $\sigma(-1)$ imply that $N \equiv 0 \bmod 8$ when $p \equiv \pm 1 \bmod 8$, respectively, $N \equiv 4 \bmod 8$ when $p \equiv \pm 3 \bmod 8$. The two computations for N modulo 8 imply that

$$\sigma(2) = \begin{cases} 1 & \text{if } p \equiv \pm 1 \bmod 8, \\ -1 & \text{if } p \equiv \pm 3 \bmod 8; \end{cases} \tag{15}$$

in compact form:

$$\sigma(2) = (-1)^{(p^2 - 1)/8}. \tag{16}$$

This explicit evaluation of $\sigma(2)$ is known as the *second supplement* to the law of quadratic reciprocity, the *first supplement* being the explicit evaluation–say, in the form (14)–of $\sigma(-1)$. These facts are named as such because they complement the prominent *law of quadratic reciprocity*, forthcoming as Theorem 1.22. It is apparent that the evaluation of $\sigma(2)$ lies deeper than that of $\sigma(-1)$. In what follows, we will use the simpler designation *supplementary law* for the evaluation of $\sigma(2)$ given by (15) or (16).

Example 1.18. Let $p > 3$. We count the number of solutions to the equation

$$x^3 + y^3 + z^3 - 3xyz = c, \tag{17}$$

where $c \in \mathbb{F}_p$.

We consider first the case when $c \neq 0$. Let us observe the factorization

$$x^3 + y^3 + z^3 - 3xyz = (x + y + z)(x^2 + y^2 + z^2 - xy - yz - zx)$$

$$= (x + y + z)\left(\frac{3}{2}(x^2 + y^2 + z^2) - \frac{1}{2}(x + y + z)^2\right).$$

We partition the solution set of (17) according to the value of $x + y + z$; that is, we think of the number of solutions as being given by the formula

$$N(c) = \sum_{a \in \mathbb{F}_p^*} N_a(c),$$

where $N_a(c)$ denotes the number of solutions satisfying $x + y + z = a$. We see that $N_a(c)$ is the number of solutions to the system

$$\begin{cases} x + y + z = a \\ x^2 + y^2 + z^2 = (2a^{-1}c + a^2)/3. \end{cases}$$

The change of variable $x := x + a/3$, $y := y + a/3$, $z := z + a/3$ leads to the simpler system

$$\begin{cases} x + y + z = 0 \\ x^2 + y^2 + z^2 = C, \qquad C = 2a^{-1}c/3. \end{cases}$$

By eliminating the variable z, this system reduces to the single equation $x^2 + y^2 + (x + y)^2 = C$, which can be rearranged as

$$3x^2 + (2y + x)^2 = C.$$

Since $C \neq 0$, we infer that $N_a(c) = p - \sigma(-3)$ for each $a \in \mathbb{F}_p^*$. Whence

$$N(c) = (p-1)\big(p - \sigma(-3)\big) = \begin{cases} (p-1)^2 & \text{if } p \equiv 1 \bmod 3, \\ p^2 - 1 & \text{if } p \equiv 2 \bmod 3, \end{cases}$$

whenever $c \neq 0$.

Finally, the identity $\sum_{c \in \mathbb{F}_p} N(c) = p^3$ yields the value of $N(0)$:

$$N(0) = 2p^2 - p + (p-1)^2 \sigma(-3) = \begin{cases} 3p^2 - 3p + 1 & \text{if } p \equiv 1 \bmod 3, \\ p^2 + p - 1 & \text{if } p \equiv 2 \bmod 3. \end{cases}$$

The quadratic equation $a_1 x_1^2 + \cdots + a_n x_n^2 = c$

We now turn to a generalization of Theorem 1.16.

Theorem 1.19. *Let $a_1, \ldots, a_n \in \mathbb{F}_p^*$, and $c \in \mathbb{F}_p$. Then the number of solutions to the equation*

$$a_1 x_1^2 + \cdots + a_n x_n^2 = c \tag{18}$$

is as follows:

- *if n is even,*

$$\begin{cases} p^{n-1} - \sigma\big((-1)^{n/2} a_1 \ldots a_n\big) p^{(n-2)/2} & \text{if } c \neq 0, \\ p^{n-1} + \sigma\big((-1)^{n/2} a_1 \ldots a_n\big)(p-1)p^{(n-2)/2} & \text{if } c = 0; \end{cases}$$

- *if n is odd,*

$$\begin{cases} p^{n-1} + \sigma\big((-1)^{(n-1)/2} a_1 \ldots a_n\big) \sigma(c) p^{(n-1)/2} & \text{if } c \neq 0, \\ p^{n-1} & \text{if } c = 0. \end{cases}$$

A rough reading of this intricate-looking count is that the number of solutions deviates from the expected value, p^{n-1}, by a term on the order of $p^{n/2}$.

The proof of the theorem uses the separation principle, much like the proof of Theorem 1.16. The bookkeeping of convolutions is just a bit more involved.

Proof of Theorem 1.19. Assume n is even. We write

$$N_{a_1 x_1^2 + \cdots + a_n x_n^2} = N_{a_1 x_1^2 + a_2 x_2^2} * \cdots * N_{a_{n-1} x_{n-1}^2 + a_n x_n^2}$$
$$= \big(p\mathbb{1} + \sigma(-a_1 a_2)\kappa\big) * \cdots * \big(p\mathbb{1} + \sigma(-a_{n-1} a_n)\kappa\big),$$

by using the separation principle, and formula (13). Now, we observe the following simple convolution rule for integral combinations of $\mathbb{1}$ and κ:

$$(t_1 \mathbb{1} + r_1 \kappa) * (t_2 \mathbb{1} + r_2 \kappa) = p(t_1 t_2 \mathbb{1} + r_1 r_2 \kappa).$$

We deduce that

$$N_{a_1 x_1^2 + \cdots + a_n x_n^2} = p^{n/2 - 1} \Big(p^{n/2} \mathbb{1} + \sigma\big((-1)^{n/2} a_1 \ldots a_n\big)\kappa \Big)$$
$$= p^{n-1} \mathbb{1} + \sigma\big((-1)^{n/2} a_1 \ldots a_n\big) p^{(n-2)/2} \kappa.$$

Explicitly, $N_{a_1 x_1^2 + \cdots + a_n x_n^2}(c)$ is given by

$$\begin{cases} p^{n-1} - \sigma\big((-1)^{n/2} a_1 \ldots a_n\big) p^{(n-2)/2} & \text{if } c \neq 0 \\ p^{n-1} + \sigma\big((-1)^{n/2} a_1 \ldots a_n\big)(p-1) p^{(n-2)/2} & \text{if } c = 0. \end{cases}$$

Assume n is odd. The solution count in the case $n = 1$ is known, and agrees with the claimed formulas; in what follows, we further assume that $n \geq 3$. We now use the separation principle to write

$$N_{a_1 x_1^2 + \cdots + a_n x_n^2} = N_{a_1 x_1^2 + \cdots + a_{n-1} x_{n-1}^2} * N_{a_n x_n^2}$$

which, in light of our previous computation, equals

$$\Big(p^{n-2} \mathbb{1} + \sigma\big((-1)^{(n-1)/2} a_1 \ldots a_{n-1}\big) p^{(n-3)/2} \kappa \Big) * \big(\mathbb{1} + \sigma(a_n)\sigma\big)$$
$$= p^{n-2}(\mathbb{1} * \mathbb{1}) + \sigma\big((-1)^{(n-1)/2} a_1 \ldots a_n\big) p^{(n-3)/2}(\kappa * \sigma)$$
$$= p^{n-1} \mathbb{1} + \sigma\big((-1)^{(n-1)/2} a_1 \ldots a_n\big) p^{(n-1)/2} \sigma.$$

Explicitly, $N_{a_1 x_1^2 + \cdots + a_n x_n^2}(c)$ is now given by

$$\begin{cases} p^{n-1} + \sigma\big((-1)^{(n-1)/2} a_1 \ldots a_n\big)\sigma(c) p^{(n-1)/2} & \text{if } c \neq 0, \\ p^{n-1} & \text{if } c = 0. \end{cases}$$

The verification is complete. $\qquad\square$

To illustrate the theorem, consider the following "spherical" examples.

Example 1.20. For $c \in \mathbb{F}_p$, the equation

$$x^2 + y^2 + z^2 = c \tag{19}$$

has $p^2 + \sigma(-c)p$ solutions.

Example 1.21. The number of solutions to the equation

$$x_1^2 + \cdots + x_n^2 = 1 \tag{20}$$

is

$$\begin{cases} p^{n-1} - \sigma(-1)^{n/2}\, p^{(n-2)/2} & \text{if } n \text{ is even,} \\ p^{n-1} + \sigma(-1)^{(n-1)/2}\, p^{(n-1)/2} & \text{if } n \text{ is odd.} \end{cases}$$

The latter example–counting points on finite "spheres"–has a very nice application: a proof of the quadratic reciprocity law. We discuss this in the next section.

Quadratic reciprocity

Theorem 1.22 (Quadratic Reciprocity). *Let p and q be distinct odd primes. Then*

$$(q/p) = (-1)^{(p-1)(q-1)/4}(p/q). \tag{21}$$

We temporarily relinquish our usage of the quadratic character σ as we are working with two different primes. Throughout this section, we revert to the classical Legendre symbol.

The signing appearing in (21) can be given an alternate form, according to the value of q modulo 4. When $q \equiv 1 \bmod 4$, we have $(-1)^{(p-1)(q-1)/4} = 1$, whereas for $q \equiv 3 \bmod 4$ we have $(-1)^{(p-1)(q-1)/4} = (-1)^{(p-1)/2} = (-1/p)$. Thus, (21) can be expressed in an equivalent, more explicit way as follows:

$$\begin{cases} (q/p) = (p/q) & \text{if } q \equiv 1 \bmod 4, \\ (-q/p) = (p/q) & \text{if } q \equiv 3 \bmod 4. \end{cases} \tag{22}$$

The following two examples are straightforward illustrations of quadratic reciprocity. In itself, the first example is not new–we have already proved it in Example 1.8, using a somewhat *ad hoc* approach. Here, there is a palpable ease.

Example 1.23. Let $p \neq 3$. Then

$$(-3/p) = \begin{cases} 1 & \text{if } p \equiv 1 \bmod 3, \\ -1 & \text{if } p \equiv 2 \bmod 3. \end{cases} \tag{23}$$

Indeed, by (22), we have $(-3/p) = (p/3)$. Now, if $p \equiv 1 \bmod 3$, then $(p/3) = (1/3) = 1$; if $p \equiv 2 \bmod 3$, then $(p/3) = (2/3) = -1$.

Example 1.24. Let $p \neq 5$. Then

$$(5/p) = \begin{cases} 1 & \text{if } p \equiv 1, 4 \bmod 5, \\ -1 & \text{if } p \equiv 2, 3 \bmod 5. \end{cases} \tag{24}$$

By (22), we have $(5/p) = (p/5)$, true reciprocity! As 1 and 4 are quadratic residues modulo 5, while 2 and 3 are not, the claimed formula follows.

Proof of Theorem 1.22. Consider the "unit sphere" of dimension q over $\mathbb{Z}/p\mathbb{Z}$, described by the equation

$$x_1^2 + \cdots + x_q^2 = 1.$$

Let N_q denote the number of solutions to this equation.

The cyclic permutation $(x_1, x_2, \ldots, x_q) \mapsto (x_q, x_1, \ldots, x_{q-1})$ defines a symmetry of order q for the solution set; geometrically, this symmetry is the rotation about the axis $x_1 = x_2 = \cdots = x_q$. Almost all orbits have size q. The possible exceptions come from fixed points, namely solutions of the form $(x, x, \ldots, x)$. Then x satisfies $qx^2 = 1$, which is solvable if and only if q is a square in $\mathbb{Z}/p\mathbb{Z}$. In other words, the number of fixed points is $1 + (q/p)$. The conclusion of this geometric perspective on the solution set is that

$$N_q \equiv 1 + (q/p) \pmod q.$$

On the other hand, we know from Example 1.21 that

$$N_q = p^{q-1} + (-1/p)^{(q-1)/2} p^{(q-1)/2} = p^{q-1} + (-1)^{(p-1)(q-1)/4} p^{(q-1)/2}.$$

We have $p^{q-1} \equiv 1 \bmod q$, and $p^{(q-1)/2} \equiv (p/q) \bmod q$, by Fermat's little theorem, respectively Euler's identity in $\mathbb{Z}/q\mathbb{Z}$. Thus,

$$N_q \equiv 1 + (-1)^{(p-1)(q-1)/4}(p/q) \pmod q.$$

Comparing the two perspectives on $N_q \bmod q$, we obtain that

$$(q/p) \equiv (-1)^{(p-1)(q-1)/4}(p/q) \pmod q.$$

As the two sides are ± 1-valued, equality must hold; so (21) is proved. $\square$

The next result is a more sophisticated use of quadratic reciprocity. For $p \equiv 1 \bmod 8$, the supplementary law (15) says that 2 is a square modulo p; thus $2^{(p-1)/2} \equiv 1 \bmod p$, and so $2^{(p-1)/4} \equiv \pm 1 \bmod p$. Determining the correct sign amounts to whether or not 2 is a fourth power modulo p. We prove the following quartic supplementary law.

Theorem 1.25. *Let $p \equiv 1 \bmod 8$, and write $p = A^2 + 2B^2$ where A and B are integers. Then*

$$2^{(p-1)/4} \equiv (-1)^{(A^2-1)/8} \pmod p \tag{25}$$

and so 2 is a fourth power $\bmod\, p$ *if and only if $A \equiv \pm 1 \bmod 8$.*

Proof. We have $2B^2 \equiv -A^2 \bmod p$; by raising to the (even) power $(p-1)/4$, we get $2^{(p-1)/4}B^{(p-1)/2} \equiv A^{(p-1)/2} \bmod p$. In view of Euler's identity, we deduce that

$$2^{(p-1)/4}(B/p) \equiv (A/p) \pmod p. \tag{26}$$

Next, we evaluate the Legendre symbols (A/p) and (B/p). As A and B need not be prime, we need to consider their prime decomposition. We might note that A is odd, B is even, and that A and B are relatively prime.

Let $A = \prod r_k$, where each r_k is an odd prime. The identity $p = A^2 + 2B^2$, viewed modulo r_k, implies that $(p/r_k) = (2/r_k)$. On the one hand, $(p/r_k) = (r_k/p)$ by quadratic reciprocity. On the other hand, the supplementary law (15) expresses $(2/r_k)$ explicitly in terms of r_k. We obtain

$$(r_k/p) = (-1)^{(r_k^2-1)/8}$$

for each k, whence

$$(A/p) = \prod (r_k/p) = (-1)^{\sum (r_k^2-1)/8}.$$

With an eye towards discarding the dependence on the r_k's by reassembling their product A, we note that

$$\frac{x^2-1}{8} + \frac{y^2-1}{8} \equiv \frac{(xy)^2-1}{8} \bmod 2$$

whenever x and y are odd integers. Indeed, by rearranging, we can rewrite the above congruence as $(x^2 - 1)(y^2 - 1) \equiv 0 \bmod 16$; this holds since $x^2, y^2 \equiv 1 \bmod 4$. We then deduce that

$$\sum \frac{r_k^2 - 1}{8} \equiv \frac{(\prod r_k)^2 - 1}{8} = \frac{A^2 - 1}{8} \quad \bmod 2$$

and so

$$(A/p) = (-1)^{(A^2-1)/8}. \tag{27}$$

Finally, we show that

$$(B/p) = 1. \tag{28}$$

The argument is similar to the one we did for A; in fact, much easier. Write $B = 2^i \prod q_j$, where each q_j is odd. Thus, $(B/p) = \prod (q_j/p)$, as $(2/p) = 1$. The identity $p = A^2 + 2B^2$, taken modulo q_j, implies that $(p/q_j) = 1$; whence $(q_j/p) = 1$ for each j, by quadratic reciprocity. Thus, $(B/p) = 1$, as desired.

By combining (27) and (28) with (26), we obtain (25). $\qquad\square$

Representing primes as sums of squares

There is a somewhat jarring aspect to Theorem 1.25: implicitly, it assumes that every prime $p \equiv 1 \bmod 8$ is of the form $p = A^2 + 2B^2$ for some integers A and B. In this section, we establish this fact, as well as two of its relatives.

Theorem 1.26. *The following hold.*

  (i) *If $p \equiv 1 \bmod 4$, then $p = A^2 + B^2$ for some integers A, B.*
  (ii) *If $p \equiv 1 \bmod 6$, then $p = A^2 + 3B^2$ for some integers A, B.*
  (iii) *If $p \equiv 1, 3 \bmod 8$, then $p = A^2 + 2B^2$ for some integers A, B.*

Let d be a positive integer which is not divisible by p, and let us consider the problem of representing

$$p = A^2 + dB^2 \tag{29}$$

for some integers A and B. Clearly, integers A and B satisfying (29) must be non-zero and relatively prime. Furthermore, if they exist then they are essentially unique, as the following lemma shows.

Lemma 1.27. *Let $p = A^2 + dB^2 = a^2 + db^2$. Then $a = \pm A$ and $b = \pm B$, or, in the case $d = 1$, $a = \pm B$ and $b = \pm A$.*

Proof. Working modulo p, we have $a^2 \equiv -db^2$ and $A^2 \equiv -dB^2$. We multiply these two congruences to deduce that $aA \equiv \pm dbB$. Up to changing the sign of b, we may assume that $aA + dbB \equiv 0$. On the other hand

$$p^2 = (a^2 + db^2)(A^2 + dB^2) = (aA + dbB)^2 + d(bA - aB)^2.$$

As $aA + dbB$ is a multiple of p, there are two possibilities. The first one is that $aA + dbB = \pm p$, in which case we must have $bA = aB$. But A and B are non-zero and relatively prime, and the same holds for a and b. A simple divisibility argument shows that $a = A$ and $b = B$, or $a = -A$ and $b = -B$.

The other possibility is that $aA + dbB = 0$. Then d divides p^2, so necessarily $d = 1$. From $aA = -bB$ one finds, as above, that $a = B$ and $b = -A$, or $a = -B$ and $b = A$. $\qquad\square$

The existence of a representation of the form (29) for p is far more subtle than its (essential) uniqueness. Firstly, we record a necessary condition on d. We continue favoring the classical Legendre symbol.

Lemma 1.28. *If $p = A^2 + dB^2$ for some integers A, B, then $(-d/p) = 1$.*

Proof. Working modulo p, we have $A^2 \equiv -dB^2$. Since B is not a multiple of p, it has a modular inverse B'–an integer satisfying $BB' \equiv 1$. Then $(B'A)^2 \equiv -d$ and so $-d$ is a quadratic residue modulo p. $\qquad\square$

Let us understand what this lemma says for the values $d = 1, 3, 2$ which appear in Theorem 1.26. For each of these values we can turn the condition $(-d/p) = 1$ into an explicit constraint on p. To have $(-1/p) = 1$ amounts to $p \equiv 1 \bmod 4$, by Example 1.7. To have $(-3/p) = 1$ amounts to $p \equiv 1 \bmod 3$, by Example 1.23; as p is odd, an alternate form is $p \equiv 1 \bmod 6$. We add that, for $d = 3$, the exceptional case $p = 3$ has to be taken into account as well. Finally, a combination of the supplementary law (15) with Example 1.7 allows us to express the condition $(-2/p) = 1$ as follows: either $(-1/p) = (2/p) = 1$, which happens precisely when $p \equiv 1 \bmod 8$, or $(-1/p) = (2/p) = -1$, which happens precisely when $p \equiv 3 \bmod 8$. To summarize:

- a representation $p = A^2 + B^2$ may exist only for $p \equiv 1 \bmod 4$;

- a representation $p = A^2 + 3B^2$ may exist only for $p \equiv 1 \bmod 6$, or $p = 3$;

- a representation $p = A^2 + 2B^2$ may exist only for $p \equiv 1, 3 \bmod 8$.

We now return to the general representation (29), and we show that the necessary condition $(-d/p) = 1$ is sufficient in a certain weak sense.

Lemma 1.29. *Assume $(-d/p) = 1$. Then there exist integers A and B such that $A^2 + dB^2 = kp$ for some integer $k \in \{1, \ldots, d\}$.*

Proof. As $(-d/p) = 1$, there is an integer m so that $m^2 \equiv -d \bmod p$.

Consider the pairs (x, y) whose entries are integral and constrained by $0 \le x, y < \sqrt{p}$. There are $\lceil \sqrt{p} \rceil$ integers in the range $[0, \sqrt{p})$, so there are more than p such pairs (x, y). Consequently, the linear form $x - my$ must have repeated values modulo p as (x, y) runs over the pairs. Let (x, y) and (x', y') be two distinct pairs satisfying $x - my \equiv x' - my' \bmod p$. Set $A = x - x'$ and $B = y - y'$. Then A and B are integers, not both zero, satisfying $|A|, |B| < \sqrt{p}$ and $A \equiv mB \bmod p$.

Now, $A^2 \equiv m^2 B^2 \equiv -dB^2 \bmod p$, and so

$$A^2 + dB^2 = kp$$

for some integer k. The bounds $0 < A^2 + dB^2 < p + dp = (d+1)p$ imply that $1 \le k \le d$. $\qquad\square$

Since we are interested in very small values of d, the previous lemma turns out to be sufficient for our purposes.

Proof of Theorem 1.26. (i) Assume that $p \equiv 1 \bmod 4$, so $(-1/p) = 1$. By using Lemma 1.28 for $d = 1$, we get $A^2 + B^2 = p$ for some integers A and B.

(ii) Assume that $p \equiv 1 \bmod 6$, so $(-3/p) = 1$. We apply Lemma 1.28 for $d = 3$. We get $A^2 + 3B^2 = kp$ for some integers A and B, where $k \in \{1, 2, 3\}$.

If $k = 1$ we are done. If $k = 3$ then the relation $A^2 + 3B^2 = 3p$ forces A to be a multiple of 3, say $A = 3A'$; therefore $3A'^2 + B^2 = p$, and we are done once again. The case $k = 2$ can be ruled out, since $A^2 + 3B^2 = 2p$ would imply $A^2 \equiv 2 \bmod 3$, a contradiction.

(iii) Assume that $p \equiv 1, 3 \bmod 8$, so $(-2/p) = 1$. By using Lemma 1.28 for $d = 2$, we get $A^2 + 2B^2 = kp$ for some integers A and B, where $k \in \{1, 2\}$.

If $k = 1$ we are done. If $k = 2$ then $A^2 + 2B^2 = 2p$ forces A to be even, say $A = 2A'$; then $2A'^2 + B^2 = p$, and we are done in this case as well. $\quad\square$

Exercises

Exercise 1.30. Let $p = 409$. Find a square root of $-1 \bmod p$, and a square root of $2 \bmod p$. Is 2 a fourth power mod p?

Exercise 1.31. Let $t \in \mathbb{F}_p$. How many matrices in $\mathrm{SL}_2(\mathbb{F}_p)$ have trace t?

Exercise 1.32. Consider the matrix $Q = (Q_{ab})_{1 \le a,b \le p}$ whose entries are defined by the following rule involving the Legendre symbol:

$$Q_{ab} = \left(\frac{a-b}{p} \right).$$

Thus, Q is a matrix of order p having 1 or -1 as off-diagonal entries, respectively, 0 along the diagonal.

(i) Show that the transpose of Q satisfies $Q^{\mathsf{T}} = (-1/p)Q$, so Q is symmetric when $p \equiv 1 \bmod 4$, respectively, antisymmetric when $p \equiv 3 \bmod 4$.
(ii) Show that $Q^2 = (-1/p) \cdot (pI_p - J_p)$, where I_p is the identity matrix of order p, and J_p is the all-1 matrix of order p.

Exercise 1.33. An $n \times n$ real matrix H is a *Hadamard matrix* if its entries are 1 or -1, and its rows are mutually orthogonal. The row orthogonality can be concisely expressed as $HH^{\mathsf{T}} = nI_n$. As $H^{\mathsf{T}}H = nI_n$, the columns of H are mutually orthogonal as well.

The following two items present constructions of Hadamard matrices. Both involve the matrix Q defined in the previous exercise.

(i) Let p be a prime satisfying $p \equiv 3 \bmod 4$. Consider the following matrix of order $p+1$:

$$H = \begin{pmatrix} 1 & 1 & \cdots & 1 \\ 1 & & & \\ \vdots & & Q - I_p & \\ 1 & & & \end{pmatrix}.$$

Show that H is a Hadamard matrix.
(ii) Let p be a prime satisfying $p \equiv 1 \bmod 4$. Consider the matrices M_+ and M_- of order $p+1$, defined by

$$M_\pm = \begin{pmatrix} \pm 1 & 1 & \cdots & 1 \\ 1 & & & \\ \vdots & & Q \pm I_p & \\ 1 & & & \end{pmatrix}.$$

Next, consider the following matrix of order $2(p+1)$:

$$\tilde{H} = \begin{pmatrix} M_+ & M_- \\ M_- & -M_+ \end{pmatrix}.$$

Show that $\tilde{H}$ is a Hadamard matrix.

Exercise 1.34. Let $a, b \in \mathbb{F}_p$. Count the number of solutions to the system of equations

$$\begin{cases} x^2 + y^2 + z^2 + t^2 = a, \\ x + y + z + t = b. \end{cases}$$

Exercise 1.35. Let $n \geq 2$. When does an n-dimensional "sphere" over $\mathbb{F}_p$, described by an equation $x_1^2 + \cdots + x_n^2 = c$, enjoy the following slice uniformity: the number of points where the sphere meets a hyperplane, that is to say, an $(n-1)$-dimensional subspace of $\mathbb{F}_p^n$, does not depend on the choice of hyperplane.

Exercise 1.36. Let $p \neq 7$. Show that $p = A^2 + 7B^2$ for some integers A and B if and only if $p \equiv 1, 2, 4 \bmod 7$.

Notes

- The existence of primitive roots mod p, in other words, the fact that the multiplicative group $\mathbb{F}_p^*$ is cyclic, is proved by Gauss in his epochal *Disquisitiones Arithmeticae* (1801). For this reason, we usually denote a generator of $\mathbb{F}_p$ by g.

  Present day mathematics is indebted to Gauss for the dictionary of modular arithmetic–terms like congruence, modulus, residues–, its framework, as well as some fundamental results–existence of primitive roots, the law of quadratic reciprocity. But such a historical viewpoint would be incomplete without an awareness of Gauss's great predecessors: Fermat, Euler, Legendre. Take, for example, the modern notation for congruence that Gauss introduced. In a footnote, Gauss acknowledges that it is a refinement of Legendre's use of equality for the congruence relation–a usage which, in retrospect, strikes us as very modern.

- Theorem 1.19 is due to Jordan (*Sur les congruences du second degré*, C.R. Acad. Sci. Paris 1866). The solution count obtained in Theorem 1.19 also shows that the number of solutions to $a_1 x_1^2 + \cdots + a_n x_n^2 = c$ is divisible by

$p^{\lceil n/2 \rceil - 1}$. A far-reaching generalization of this fact is the following result due to Ax (*Zeroes of polynomials over finite fields*, Amer. J. Math. 1964): the number of solutions to a polynomial equation $f(x_1, \ldots, x_n) = c$ over $\mathbb{F}_p$ is divisible by $p^{\lceil n/d \rceil - 1}$, where d is the total degree of f.

- Quadratic reciprocity is the heirloom of early number theorists: Euler, Legendre, Gauss. For Gauss, in particular, it was the gem of higher arithmetic, his *theorema aureum*–the golden theorem. In his quest for cubic and quartic reciprocity laws, Gauss returned to the law of quadratic reciprocity and gave several different proofs over a period of nearly twenty years. The first chapter of Lemmermeyer's monograph *Reciprocity laws. From Euler to Eisenstein* (Springer 2000) gives a highly readable account of the history.

  The solution-counting proof for the law of quadratic reciprocity, as well as that for the supplementary law (Example 1.17), are due to Victor-Amédée Lebesgue (*Recherches sur les nombres*, J. de Math. Pures Appl. 1838).

- There are two possible values for $2^{(p-1)/2} \bmod p$, namely 1 or -1; the supplementary law explicitly determines which one occurs for a given p. When $p \equiv 1 \bmod 4$, there are four possible values for $2^{(p-1)/4} \bmod p$, namely $1, -1, j$, and $-j$, where j stands for an integer satisfying $j^2 \equiv -1 \bmod p$. A quartic supplementary law, pinpointing the correct outcome, is possible but its statement is somewhat complicated. Theorem 1.25 offers a simplified quartic supplementary law for the case $p \equiv 1 \bmod 8$; this version is originally due to Gauss. The proof of Theorem 1.25 could be streamlined by introducing the Jacobi symbol, a natural extension of the Legendre symbol to arithmetic modulo n where n is not necessarily prime.

- The problem of writing a given prime p as a sum of (repeated) squares, $p = A^2 + dB^2$ for some integers A and B, already appears in the work of Diophantus (3rd century AD). But the systematic study of which primes admit such sum-of-squares representations starts with Fermat. His most famous result in this vein is

  (Fer$_4$) a prime $p \equiv 1 \bmod 4$ admits a representation $p = A^2 + B^2$.

  Actually, Albert Girard had stated this result, in print, ahead of Fermat. But Fermat also obtained further results of this type:

  (Fer$_6$) a prime $p \equiv 1 \bmod 6$ admits a representation $p = A^2 + 3B^2$;
  (Fer$_8$) a prime $p \equiv 1, 3 \bmod 8$ admits a representation $p = A^2 + 2B^2$.

Fermat did not document his proofs, though he did suggest that he got to (Fer$_4$) by his method of infinite descent. About a century after Fermat, Euler managed to prove the above results. Other claims of Fermat resisted Euler's repeated attempts–notably, the four-squares theorem stating every prime admits an integral representation of the form $A^2 + B^2 + C^2 + D^2$. There is also a "theorem" of Fermat that Euler actually disproved: the statement that all Fermat numbers are primes. At any rate, Theorem 1.26 is commonly ascribed to Fermat.

The elementary proof of Theorem 1.26 given herein appears in several classical books on number theory, such as the one by Nagell (*Introduction to Number Theory*, second edition, Chelsea Publishing Co. 1964). The pigeon-hole argument used in the proof of Lemma 1.29 is an idea due to Axel Thue.

The claims of Theorem 1.26 are existence statements, and most proofs provide just that. This is the case with the proof we have presented. Fermat's proofs were presumably applications of his "method of descent", which is essentially an argument by contradiction. Heath-Brown's fixed-point proof of (Fer$_4$), provocatively distilled by Zagier to one sentence (*A one-sentence proof that every prime $p \equiv 1 \pmod 4$ is a sum of two squares*, Amer. Math. Monthly 1990), is an existential argument, as well.

Explicit formulas for A and B in (Fer$_4$) were first provided by Gauss, in terms of binomial coefficients. In Chapter 3, we will obtain explicit formulas for A and B in terms of Jacobsthal sums, in the cases $p \equiv 1$ mod 4, $p \equiv 1$ mod 6, and $p \equiv 1$ mod 8.

There is an algorithmic approach to representing primes in the form $A^2 + dB^2$; see Brillhart (*Note on representing a prime as a sum of two squares*, Math. Comp. 1972) and Basilla (On the solution of $x^2 + dy^2 = m$, *Proc. Japan Acad. Ser. A Math.*, 2004). For much more about the fascinating problem of representing primes in the form $A^2 + dB^2$, we refer to the beautiful account by Cox (*Primes of the form $x^2 + ny^2$. Fermat, class field theory, and complex multiplication*, Second edition, John Wiley & Sons 2013).

- A Hadamard matrix of order $n > 2$ can exist only when $n \equiv 0$ mod 4. Indeed, up to multiplying each column by -1, we may assume that the first row of an $n \times n$ Hadamard matrix consists entirely of 1's. Next, we consider the second and the third row, and we count the possible patterns as follows: denote by n_+^+, n_-^+, n_+^-, and n_-^- the number of pairs of the form

$\binom{1}{1}$, $\binom{1}{-1}$, $\binom{-1}{1}$, and $\binom{-1}{-1}$, respectively. The following relations reflect the mutual orthogonality between the first three rows:

$$n_+^+ - n_-^+ - n_+^- + n_-^- = 0,$$

$$n_+^+ + n_-^+ - n_+^- - n_-^- = 0,$$

$$n_+^+ - n_-^+ + n_+^- - n_-^- = 0.$$

The upshot is that $n_+^+ = n_-^+ = n_+^- = n_-^-$. As $n_+^+ + n_-^+ + n_+^- + n_-^- = n$, it follows that $n \equiv 0 \bmod 4$.

Does there exist a Hadamard matrix of order n for every $n \equiv 0 \bmod 4$? This simple-minded question is still unresolved, though many positive results are known by now. Exercise 1.33 highlights a remarkable method for constructing Hadamard matrices of order $n = p+1$, for $p \equiv 3 \bmod 4$, respectively of order $n = 2(p + 1)$, for $p \equiv 1 \bmod 4$. The method is due to Paley (*On orthogonal matrices*, J. Math. Phys. 1933). At the heart of Paley's construction is the summation formula of Theorem 1.11. This is a result due, to the best of our knowledge, to Ernst Jacobsthal, whose name will recur throughout this text. Interestingly, Paley does not reference Jacobsthal in his paper.

The Paley graphs defined in Example 1.12 are fundamental examples in algebraic graph theory. Paley never defined them; the name owes to the fact that their construction is related to Paley's construction of Hadamard matrices. For more on Paley graphs–their origin, and their properties–we refer to the nice account by Jones (*Paley and the Paley graphs*, in "Isomorphisms, Symmetry and Computations in Algebraic Graph Theory", Springer 2020).

Chapter 2

Quadratic Character Sums

Introducing quadratic character sums

A *quadratic character sum* is a sum of the form

$$\sum_{x \in \mathbb{F}_p} \sigma(f(x)), \tag{30}$$

where f is a polynomial map with coefficients in $\mathbb{F}_p$. When f is linear, the sum (30) vanishes; for f permutes the elements of $\mathbb{F}_p$ and so the sum is really (5) in disguise. When f is quadratic, the sum (30) can still be evaluated; the answer turns out to be simple and satisfactory, though the argument is not quite as evident as in the linear case; this is discussed in the next section. When f is cubic, the explicit evaluation of (30) is already intractable in general. Later parts of this text are concerned with quadratic character sums whose polynomial arguments have degree 3 or higher.

We get started with some general observations on quadratic character sums. The first one highlights a fundamental link to solution counting.

Lemma 2.1. *The number of solutions to the equation $y^2 = f(x)$ is*

$$p + \sum_{x \in \mathbb{F}_p} \sigma(f(x)).$$

Proof. For each $x \in \mathbb{F}_p$, the equation $y^2 = f(x)$ has $1 + \sigma(f(x))$ solutions. The solution count is therefore

$$\sum_{x \in \mathbb{F}_p} \left(1 + \sigma(f(x))\right) = p + \sum_{x \in \mathbb{F}_p} \sigma(f(x))$$

as claimed. $\qquad\square$

25

Lemma 2.2. *The quadratic character sum*

$$\sum_{x \in \mathbb{F}_p} \sigma(f(x))$$

is an integer in the interval $[-p, p]$*, whose parity is opposite to that of the number of distinct roots that* f *has in* $\mathbb{F}_p$*.*

Proof. Each term in the quadratic character sum is 0, 1, or -1, so the sum is an integer between $-p$ and p inclusive. We have $\sigma(f(x)) = 0$ whenever x is a root of f, and $\sigma(f(x)) \equiv 1 \bmod 2$ otherwise. Let n_0 denote the number of distinct roots that f has in $\mathbb{F}_p$. Then

$$\sum_{x \in \mathbb{F}_p} \sigma(f(x)) \equiv p - n_0 \bmod 2.$$

As p is odd, the parity of $p - n_0$ is opposite to that of n_0. $\qquad\square$

The next observation is a useful trick for lowering the degree of the polynomial argument in a quadratic character sum.

Lemma 2.3. *We have*

$$\sum_{x \in \mathbb{F}_p} \sigma(f(x^2)) = \sum_{x \in \mathbb{F}_p} \sigma(f(x)) + \sum_{x \in \mathbb{F}_p} \sigma(xf(x)). \tag{31}$$

Proof. Indeed, we have

$$\sum_{x \in \mathbb{F}_p} \sigma(f(x^2)) = \sum_{y \in \mathbb{F}_p} \sigma(f(y)) \cdot \#\{x \in \mathbb{F}_p : y = x^2\}$$

$$= \sum_{y \in \mathbb{F}_p} \sigma(f(y))\big(1 + \sigma(y)\big).$$

We obtain (31) by separating into two sums, and reverting to the original variable x. $\qquad\square$

We will refer to (31) as the *downgrading formula*. On the left-hand side, the polynomial argument has degree $2\deg(f)$; on the right-hand side, the polynomial arguments have degree $\deg(f)$, respectively, $\deg(f) + 1$.

Quadratic character sums with quadratic arguments

In this section, we evaluate the quadratic character sum (30) in the case when f is a quadratic polynomial. Such sums occur frequently in the calculus of quadratic character sums.

Theorem 2.4. *Let $b, c \in \mathbb{F}_p$. We have*

$$\sum_{x \in \mathbb{F}_p} \sigma(x^2 + bx + c) = \begin{cases} -1 & \text{if } b^2 - 4c \neq 0, \\ p - 1 & \text{if } b^2 - 4c = 0. \end{cases} \tag{32}$$

We give three proofs, in order to introduce basic techniques that will be used later on. When needed, we use the familiar notation $\Delta = b^2 - 4c$.

First Proof. We first handle the simpler case when $b = 0$. By the downgrading formula (31), we can write

$$\sum_{x \in \mathbb{F}_p} \sigma(x^2 + c) = \sum_{x \in \mathbb{F}_p} \sigma(x + c) + \sum_{x \in \mathbb{F}_p} \sigma(x(x + c)).$$

On the right-hand side, the first sum vanishes thanks to (5). We evaluate the second sum by using Theorem 1.11:

$$\sum_{x \in \mathbb{F}_p} \sigma(x(x + c)) = \sigma(-1) \sum_{x \in \mathbb{F}_p} \sigma(-x)\sigma(x + c) = \begin{cases} -1 & \text{if } c \neq 0, \\ p - 1 & \text{if } c = 0. \end{cases}$$

The desired formula is thereby verified when $b = 0$. Next, we deduce the general result by completing the square:

$$x^2 + bx + c = \left(x + \frac{b}{2}\right)^2 - \frac{\Delta}{4}.$$

Thus, after a change of variable $x := x - (b/2)$, we obtain

$$\sum_{x \in \mathbb{F}_p} \sigma(x^2 + bx + c) = \sum_{x \in \mathbb{F}_p} \sigma\left(x^2 - \frac{\Delta}{4}\right) = \begin{cases} -1 & \text{if } \Delta \neq 0, \\ p - 1 & \text{if } \Delta = 0. \end{cases} \qquad \square$$

Second Proof. As in the previous proof, it suffices to handle the case $b = 0$. Put

$$S(c) = \sum_{x \in \mathbb{F}_p} \sigma(x^2 + c).$$

For $c = 0$, we can easily evaluate $S(0) = \sum_{x \in \mathbb{F}_p} \sigma(x^2) = p - 1$. Our aim is to show that $S(c) = -1$ for all $c \in \mathbb{F}_p^*$.

Firstly, observe that the value of $S(c)$ only depends on whether c is a square or not. Indeed, the change of variable $x := sx$, where $s \neq 0$, gives

$$S(c) = \sum_{x \in \mathbb{F}_p} \sigma(x^2 + c) = \sum_{x \in \mathbb{F}_p} \sigma(s^2 x^2 + c) = \sum_{x \in \mathbb{F}_p} \sigma(x^2 + cs^{-2}) = S(cs^{-2}).$$

We put $S(c) = S_+$ when $\sigma(c) = 1$, respectively, $S(c) = S_-$ when $\sigma(c) = -1$.

Secondly, we work out another change of variable in the sum $S(c)$, namely the inversion $x := x^{-1}$. To do so, we need to take aside the term in $S(c)$ corresponding to $x = 0$. We have

$$\begin{aligned}
S(c) - \sigma(c) &= \sum_{x \in \mathbb{F}_p^*} \sigma(x^2 + c) = \sum_{x \in \mathbb{F}_p^*} \sigma(x^{-2} + c) \\
&= \sum_{x \in \mathbb{F}_p^*} \sigma(1 + cx^2) = \sigma(c) \sum_{x \in \mathbb{F}_p^*} \sigma(x^2 + c^{-1}) \\
&= \sigma(c)\big(S(c^{-1}) - \sigma(c^{-1})\big) = \sigma(c)S(c^{-1}) - 1.
\end{aligned}$$

This relation carries no new content when c is a square in $\mathbb{F}_p^*$. But when c is a non-square, it gives $S_- = -1$.

Thirdly, and finally, we consider the global sum

$$\sum_{c \in \mathbb{F}_p} S(c) = \sum_{x \in \mathbb{F}_p} \sum_{c \in \mathbb{F}_p} \sigma(x^2 + c).$$

The right-hand side is 0, since each inner sum vanishes thanks to (5). On the left-hand side, the term corresponding to $c = 0$ contributes $S(0) = p - 1$, while the terms corresponding to non-zero values of c contribute $\frac{1}{2}(p-1)S_+ + \frac{1}{2}(p-1)S_-$. Hence,

$$(p - 1) + \frac{p - 1}{2} S_+ + \frac{p - 1}{2} S_- = 0,$$

and so $S_+ + S_- = -2$. Since we already know that $S_- = -1$, we find that $S_+ = -1$ as well. We conclude that $S(c) = -1$ for all $c \in \mathbb{F}_p^*$, as desired. $\qquad\square$

Third Proof. Put

$$S = \sum_{x \in \mathbb{F}_p} \sigma(x^2 + bx + c),$$

an integer in the interval $[-p, p]$. In view of Euler's identity, we have

$$S \equiv \sum_{x \in \mathbb{F}_p} (x^2 + bx + c)^{(p-1)/2} \bmod p.$$

We evaluate the right-hand sum by using Corollary 1.2. Since the polynomial map $(x^2 + bx + c)^{(p-1)/2}$ has degree $p-1$ and leading coefficient 1, the above sum equals -1 in $\mathbb{F}_p$. Thus, $S \equiv -1 \bmod p$. At this point, we know that

$$S = -1 \text{ or } S = p - 1.$$

In order to discern between the two possibilities, we only need to know the parity of S. To that end, we invoke Lemma 2.2. When $\Delta \neq 0$, the quadratic argument $x^2 + bx + c$ has 0 or 2 distinct roots–at any rate, an even number of roots; in this case S is odd, so $S = -1$. When $\Delta = 0$, the quadratic argument $x^2 + bx + c$ has a single root; in this case S is even, so $S = p - 1$. $\qquad\square$

For ease of reference, we record some specific instances of Theorem 2.4 that are commonly encountered. It will be convenient to start using the *Iverson bracket* notation: if P is a statement, then

$$[\![P]\!] = \begin{cases} 1 & \text{if } P \text{ is true,} \\ 0 & \text{if } P \text{ is false.} \end{cases} \tag{33}$$

We then have

$$\sum_{x \in \mathbb{F}_p} \sigma(x^2 + c) = p[\![c = 0]\!] - 1, \tag{34}$$

$$\sum_{x \in \mathbb{F}_p} \sigma(x^2 + cx) = p[\![c = 0]\!] - 1, \tag{35}$$

$$\sum_{x \in \mathbb{F}_p} \sigma\big((x + c)(x + c')\big) = p[\![c = c']\!] - 1. \tag{36}$$

Let us now turn to applications of the above formulas. We start with the following example.

Example 2.5. Which event is more likely in $\mathbb{F}_p$: a run of two consecutive squares, or a run of two consecutive non-squares?

Having fixed a quadratic signature $\epsilon = \pm 1$, we wish to compute

$$N_\epsilon = \#\{x \in \mathbb{F}_p : \sigma(x) = \sigma(x+1) = \epsilon\}.$$

For $x \in \mathbb{F}_p$, the product $(\sigma(x) + \epsilon)(\sigma(x+1) + \epsilon)$ equals 4 when $\sigma(x) = \sigma(x+1) = \epsilon$; or $1 + \epsilon$ when $x = 0$; or $1 + \epsilon\sigma(-1)$ when $x = -1$; or 0 in all other cases. Thus,

$$\sum_{x \in \mathbb{F}_p} (\sigma(x) + \epsilon)(\sigma(x+1) + \epsilon) = 4N_\epsilon + 2 + \epsilon + \epsilon\sigma(-1).$$

On the other hand, we can compute the left-hand side by expanding:

$$\sum_{x \in \mathbb{F}_p} \sigma(x(x+1)) + \epsilon \sum_{x \in \mathbb{F}_p} \sigma(x) + \epsilon \sum_{x \in \mathbb{F}_p} \sigma(x+1) + \sum_{x \in \mathbb{F}_p} 1 = -1 + p,$$

since the first sum equals -1, by (35), while the next two sums vanish, by (5). Therefore,

$$N_\epsilon = \begin{cases} (p-3)/4 & \text{if } p \equiv 3 \bmod 4, \\ (p-1)/4 - (\epsilon+1)/2 & \text{if } p \equiv 1 \bmod 4. \end{cases}$$

The answer to our initial query is as follows: when $p \equiv 3 \bmod 4$, it is equally likely to have two consecutive squares or two consecutive non-squares; when $p \equiv 1 \bmod 4$, it is slightly less likely to have two consecutive squares than two consecutive non-squares.

Applications to solution counting

Example 2.6. We revisit the quadratic equation $ax^2 + by^2 = c$, where $a, b \in \mathbb{F}_p^*$ and $c \in \mathbb{F}_p$. The equation can be rewritten as $y^2 = b^{-1}(-ax^2 + c)$; by Lemma 2.1, the number of solutions is

$$p + \sum_{x \in \mathbb{F}_p} \sigma\big(b^{-1}(-ax^2 + c)\big) = p + \sigma(-ab) \sum_{x \in \mathbb{F}_p} \sigma\big(x^2 - a^{-1}c\big)$$

$$= p + \sigma(-ab)\big(p[\![c = 0]\!] - 1\big).$$

We recover the count of Theorem 1.16.

Example 2.7. Let $c \in \mathbb{F}_p$. We count the number of solutions to the equation

$$x^2 + y^2 + z^2 = xyz + c. \tag{37}$$

For each $x, y \in \mathbb{F}_p$, (37) is a quadratic equation in z, namely $z^2 - xyz + x^2 + y^2 - c = 0$; this equation has $1 + \sigma\big(x^2y^2 - 4(x^2 + y^2 - c)\big)$ solutions. Thus, the number of solutions to (37) is

$$N(c) = \sum_{x,y \in \mathbb{F}_p} \big(1 + \sigma\big(x^2y^2 - 4(x^2 + y^2 - c)\big)\big)$$

$$= p^2 + \sum_{x \in \mathbb{F}_p} \sum_{y \in \mathbb{F}_p} \sigma\big((x^2 - 4)y^2 - 4(x^2 - c)\big).$$

When $x^2 = 4$, the inner sum equals $\sigma(-4(x^2 - c))p = \sigma(c - 4)p$. There are two such values of x; their contribution to $N(c)$ is thus $2\sigma(c - 4)p$. When $x^2 \neq 4$, the inner sum equals $\sigma(x^2 - 4)(p[\![x^2 = c]\!] - 1)$ by (34). The contribution of these terms to $N(c)$ is

$$\sum_{x \in \mathbb{F}_p} \sigma(x^2 - 4)(p[\![x^2 = c]\!] - 1) = 1 + p \sum_{x \in \mathbb{F}_p} \sigma(x^2 - 4)[\![x^2 = c]\!].$$

In the left-hand sum, we have included the terms with $x^2 = 4$, as they are harmless. The sum on the right-hand side has at most two terms; we easily see that it equals $(1 + \sigma(c))\sigma(c - 4)$. To summarize, we have

$$N(c) = p^2 + 1 + \big(3 + \sigma(c)\big)\sigma(c - 4)p. \tag{38}$$

Let us add a few comments on the Equation (37). Over the integers, the diophantine equation $x^2 + y^2 + z^2 = 3xyz$ is known as Markov's equation. Over the finite field $\mathbb{F}_p$, Markov's equation is equivalent to

$$x^2 + y^2 + z^2 = xyz$$

as long as $p \neq 3$; quite obviously, it is the rescaling $(x, y, z) \mapsto (x/3, y/3, z/3)$ that implements the equivalence. This explains why Equation (37) for $c = 0$ is referred to as the *Markov cubic*. According to formula (38), the Markov cubic has $p^2 + 1 + 3\sigma(-1)p$ solutions over $\mathbb{F}_p$.

On the other hand, Equation (37) for $c = 4$, that is

$$x^2 + y^2 + z^2 = xyz + 4$$

is known as the *Cayley cubic*; by formula (38), it has $p^2 + 1$ solutions over $\mathbb{F}_p$.

Example 2.8. Let $a, b \in \mathbb{F}_p^*$ be distinct. We count the number of solutions to the equation

$$(x^2 + y^2)^2 = ax^2 + by^2. \tag{39}$$

In the usual Cartesian plane, a quartic curve of the form (39) is known as a *Booth curve*, or a hippopede. A distinguished example is the lemniscate, which occurs for $a = 1$ and $b = -1$.

The number of solutions to (39) is $N = \sum_{c \in \mathbb{F}_p} N(c)$, where $N(c)$ is the number of solutions to the system

$$\begin{cases} x^2 + y^2 = c, \\ ax^2 + by^2 = c^2. \end{cases}$$

As the above system amounts to

$$x^2 = \frac{c^2 - bc}{a - b}, \qquad y^2 = \frac{c^2 - ac}{b - a},$$

we quickly see that

$$N(c) = \left(1 + \sigma\left(\frac{c^2 - bc}{a - b}\right)\right)\left(1 + \sigma\left(\frac{c^2 - ac}{b - a}\right)\right).$$

We can now evaluate

$$N = \sum_{c \in \mathbb{F}_p}\left(1 + \sigma\left(\frac{c^2 - bc}{a - b}\right)\right)\left(1 + \sigma\left(\frac{c^2 - ac}{b - a}\right)\right).$$

We have

$$\sum_{c \in \mathbb{F}_p} \sigma(c^2 - bc) = \sum_{c \in \mathbb{F}_p} \sigma(c^2 - ac) = -1$$

by (35), respectively,

$$\sum_{c \in \mathbb{F}_p} \sigma\left((c^2 - bc)(c^2 - ac)\right) = \sum_{c \in \mathbb{F}_p^*} \sigma\left((c - b)(c - a)\right) = -1 - \sigma(ab)$$

by (36). Therefore,

$$N = p - \sigma(a - b) - \sigma(b - a) - \sigma(-1)\left(1 + \sigma(ab)\right).$$

A concrete example is the case $a = 1$, $b = -1$; this is the lemniscate over $\mathbb{F}_p$. The number of solutions takes the form $p - \left(1 + \sigma(-1)\right)\left(1 + \sigma(2)\right)$, which evaluates as $p - 4$ when $p \equiv 1 \bmod 8$, respectively, p otherwise.

Example 2.9. We count the number of solutions to the equation

$$\left(1 + x + \frac{1}{x}\right)\left(1 + y + \frac{1}{y}\right) = 1. \tag{40}$$

For each $c \in \mathbb{F}_p^*$, the equations $1 + x + x^{-1} = c$, $1 + y + y^{-1} = c^{-1}$ have $1 + \sigma((c-1)^2 - 4)$, respectively $1 + \sigma((c^{-1} - 1)^2 - 4)$ solutions. So, the number of solutions to Equation (40) is

$$N = \sum_{c \in \mathbb{F}_p^*} \left(1 + \sigma\big((c-1)^2 - 4\big)\right)\left(1 + \sigma\big((c^{-1} - 1)^2 - 4\big)\right).$$

Note that

$$\sum_{c \in \mathbb{F}_p^*} \sigma\big((c^{-1} - 1)^2 - 4\big) = \sum_{c \in \mathbb{F}_p^*} \sigma\big((c - 1)^2 - 4\big) = -1 - \sigma(-3)$$

by (34). Also,

$$\sigma\big((c^{-1} - 1)^2 - 4\big)\sigma\big((c - 1)^2 - 4\big) = \sigma\big((c^{-1} + 1)(c^{-1} - 3)(c + 1)(c - 3)\big)$$
$$= \sigma\big((1 - 3c)(c - 3)\big),$$

whenever $c \neq 0, -1$. We reach the formula

$$N = (p - 1) - 2 - 2\sigma(-3) + \sum_{c \neq 0, -1} \sigma\big((1 - 3c)(c - 3)\big)$$
$$= p - 3 - 3\sigma(-3) - \sigma(-1) + \sum_{c \in \mathbb{F}_p} \sigma\big((1 - 3c)(c - 3)\big).$$

Now,

$$\sum_{c \in \mathbb{F}_p} \sigma\big((1 - 3c)(c - 3)\big) = \sigma(-3)\sum_{c \in \mathbb{F}_p} \sigma\big((c - 3^{-1})(c - 3)\big) = -\sigma(-3),$$

by (36). Actually, the above calculation only works when $p \neq 3$; however, its outcome remains valid when $p = 3$ as well.

We conclude that the number of solutions to Equation (40) is

$$N = p - 3 - 4\sigma(-3) - \sigma(-1).$$

Values of cubic maps

A quadratic map on $\mathbb{F}_p$ takes exactly $(p+1)/2$ values. Indeed, consider a map $f(x) = ax^2 + bx + c$, where $a \neq 0$. We can rewrite f as $f(x) = a(x+b')^2 + c'$, for suitable b' and c'. By translating and scaling the image of f, operations which preserve its size, we are down to the simplest quadratic map, $x \mapsto x^2$. As we know, this map takes exactly $(p+1)/2$ distinct values.

Next, we determine the number of distinct values for cubic maps.

Theorem 2.10. *Let $p > 3$. The number of values of a cubic map on $\mathbb{F}_p$, of the form $f(x) = ax^3 + bx^2 + cx + d$ with $a \neq 0$, is given as follows:*

- *if $b^2 = 3ac$, then*

$$\#f(\mathbb{F}_p) = \begin{cases} (p+2)/3 & \text{if } p \equiv 1 \bmod 3, \\ p & \text{if } p \equiv 2 \bmod 3, \end{cases} \tag{41}$$

- *if $b^2 \neq 3ac$, then*

$$\#f(\mathbb{F}_p) = \begin{cases} (2p+1)/3 & \text{if } p \equiv 1 \bmod 3, \\ (2p-1)/3 & \text{if } p \equiv 2 \bmod 3. \end{cases} \tag{42}$$

Proof. As we did above for quadratics, we begin by trading the general cubic f for a much simpler cubic, without changing the size of the image $f(\mathbb{F}_p)$. We scale f by a, so as to replace $f(x) = ax^3 + bx^2 + cx + d$ by $f(x) = x^3 + (b/a)x^2 + (c/a)x + (d/a)$. We change the variable $x := x - b/(3a)$, so as to replace $f(x) = x^3 + (b/a)x^2 + (c/a)x + (d/a)$ by its depressed form $f(x) = x^3 + \alpha x + \beta$. Finally, we ranslating f by β so as to get

$$g(x) = x^3 + \alpha x, \qquad \alpha = \frac{3ac - b^2}{3a^2}.$$

Assume that $\alpha = 0$, so $g(x) = x^3$. If $p \equiv 2 \bmod 3$ then g is bijective on $\mathbb{F}_p$, so $\#g(\mathbb{F}_p) = p$ in this case. If $p \equiv 1 \bmod 3$ then $\#g(\mathbb{F}_p^*) = (p-1)/3$, so $\#g(\mathbb{F}_p) = (p+2)/3$. Formula (41) is thereby checked.

Now assume that $\alpha \neq 0$. For $k = 1, 2, 3$, consider the multiplicity count m_k, defined as the number of $x \in \mathbb{F}_p$ with the property that $g(y) = g(x)$ has k distinct solutions as an equation in y. The number of distinct values taken on by g is then given by

$$\#g(\mathbb{F}_p) = m_1 + \frac{m_2}{2} + \frac{m_3}{3}.$$

We also have $m_1 + m_2 + m_3 = p$. So, in order to evaluate $\#g(\mathbb{F}_p)$, it suffices to find only two of the three multiplicities. We focus on m_1 and m_2; we will then evaluate

$$\#g(\mathbb{F}_p) = m_1 + \frac{m_2}{2} + \frac{p - m_1 - m_2}{3} = \frac{p}{3} + \frac{2m_1}{3} + \frac{m_2}{6}. \tag{43}$$

The equation $g(y) = g(x)$ reads as $y^3 + \alpha y = x^3 + \alpha x$, and so we need to count the distinct solutions to the quadratic equation in y

$$y^2 + xy + x^2 + \alpha = 0 \tag{44}$$

which are different from x. The discriminant of the Equation (44) is

$$\Delta(x) = -3x^2 - 4\alpha.$$

The multiplicity m_1 counts those $x \in \mathbb{F}_p$ for which either (44) has no solution, or (44) has exactly one solution $y = x$. Actually, the latter case cannot occur: $\Delta(x) = 0$ and $3x^2 + \alpha = 0$ are incompatible, since $\alpha \neq 0$. There are no solutions to (44) precisely when $\sigma(\Delta(x)) = -1$, so we obtain

$$m_1 = \sum_{x \in \mathbb{F}_p} [\![\sigma(\Delta(x)) = -1]\!]$$

$$= \sum_{x \in \mathbb{F}_p} \left(\frac{1 - \sigma(\Delta(x))}{2} - \frac{[\![\Delta(x) = 0]\!]}{2} \right)$$

$$= \frac{p + \sigma(-3)}{2} - \frac{1 + \sigma(-3\alpha)}{2}.$$

Indeed, $\Delta(x) = 0$ is achieved for $1 + \sigma(-3\alpha)$ values of $x \in \mathbb{F}_p$, and by (34)

$$\sum_{x \in \mathbb{F}_p} \sigma(\Delta(x)) = -\sigma(-3).$$

The multiplicity m_2 counts those $x \in \mathbb{F}_p$ for which either (44) has a unique solution, different from x, or (44) has two solutions, one of which is x. The first case occurs when $\Delta(x) = 0$; as we have already noted, the unique solution to (44) is different from x. This case contributes $1 + \sigma(-3\alpha)$ towards m_2. The second case occurs when $\sigma(\Delta(x)) = 1$ and $3x^2 + \alpha = 0$. We note that $3x^2 + \alpha = 0$ makes $\Delta(x) = 9x^2$, so $\sigma(\Delta(x)) = 1$ holds automatically. This case also contributes $1 + \sigma(-3\alpha)$ towards m_2. Thus,

$$m_2 = 2(1 + \sigma(-3\alpha)).$$

Turning now to (43), pleasant simplifications yield

$$\#g(\mathbb{F}_p) = \frac{2p + \sigma(-3)}{3}.$$

We arrive at the formula (42) by using the explicit evaluation of $\sigma(-3)$. $\qquad\square$

Let us illustrate the scope of the previous theorem on two examples–the first one straightforward, the second one less so.

Example 2.11. Let $p > 3$ and fix $a \in \mathbb{F}_p^*$. For how many choices of $b \in \mathbb{F}_p$ is the polynomial $X^3 + aX + b$ irreducible over $\mathbb{F}_p$?

For cubic polynomials, irreducibility is equivalent to the lack of roots. Thus, $X^3 + aX + b$ is irreducible precisely when $-b$ is not a value of the cubic map $x^3 + ax$ on $\mathbb{F}_p$. Theorem 2.10 implies that, irrespective of $a \in \mathbb{F}_p^*$, the number of choices for b is

$$\begin{cases} (p-1)/3 & \text{if } p \equiv 1 \bmod 3, \\ (p+1)/3 & \text{if } p \equiv 2 \bmod 3. \end{cases}$$

Example 2.12. Consider a cubic map of the form $f(x) = x^3 + bx^2 + cx$, assumed to have no repeated roots in $\mathbb{F}_p$. We discuss the possibility that

$$\sum_{x \in \mathbb{F}_p} \sigma(f(x)) \equiv -1 \bmod p. \tag{45}$$

By hypothesis, f has either one or three distinct roots in $\mathbb{F}_p$. We infer, by Lemma 2.2, that the corresponding quadratic character sum is an even integer in the interval $[-p, p]$; so it is congruent to -1 modulo p if and only if it equals $p - 1$. Since $\sigma(f(x)) = 0$ for $x = 0$, it must be that $\sigma(f(x)) = 1$ for each $x \neq 0$. Thus,

$$f(\mathbb{F}_p^*) \subseteq (\mathbb{F}_p^*)^2. \tag{$*$}$$

As we are about to see, this is a very strong restriction. Assuming $p > 3$, we appeal to Theorem 2.10. Let us analyze the two possible cases.

When $b^2 = 3c$, the first observation is that $(*)$ can only hold in the case $p \equiv 1 \bmod 3$. However, by writing

$$f(x) = x^3 + bx^2 + \frac{b^2}{3}x = \left(x + \frac{b}{3}\right)^3 - \left(\frac{b}{3}\right)^3,$$

we can make a second observation: when $p \equiv 1 \bmod 3$, the map f has two additional roots in $\mathbb{F}_p$ besides $x = 0$ since the equation $z^3 = 1$ has three solutions in $\mathbb{F}_p$. This violates $(*)$. In short, $b^2 = 3c$ is a dead end. When $b^2 \neq 3c$, the size of $f(\mathbb{F}_p^*)$ is on the order of $2p/3$, whereas the size of $(\mathbb{F}_p^*)^2$ is on the order of $p/2$; so, for $(*)$ to hold, p needs to be very small. More precisely,

$$\frac{p-1}{2} = \#(\mathbb{F}_p^*)^2 \geq \#f(\mathbb{F}_p^*) \geq \frac{2p-1}{3} - 1,$$

which can only hold for $p \leq 5$. We have thus narrowed in the existence of f to only two primes, $p = 3$ and $p = 5$.

For $p = 3$, $(*)$ says that $f(x) = 1$, whenever $x \neq 0$; the system $f(1) = f(-1) = 1$ leads to $f(x) = x^3 + x^2 - x$.

For $p = 5$, $(*)$ says that $f(x) = \pm 1$, whenever $x \neq 0$. In particular, the values $f(1) = 1 + b + c$ and $f(-1) = -1 + b - c$ equal ± 1. If $f(1) = f(-1) = \pm 1$ then $c = -1$ and $b = \pm 1$, so $f(x) = x^3 \pm x^2 - x$; however, the value $f(2) = 1 \mp 1$ is not 1 or -1. Consequently, $f(1) = \pm 1$ and $f(-1) = \mp 1$; then $b = 0$ and $c + 1 = \pm 1$. The case $c = 0$ yields $f(x) = x^3$, which is disallowed by its having repeated roots. The case $c = -2$ yields $f(x) = x^3 - 2x$. This cubic has no repeated roots, and satisfies $f(\pm 1) = \mp 1$ and $f(\pm 2) = \mp 1$.

To summarize, we have found just two cubics: $f(x) = x^3 + x^2 - x$ over $\mathbb{F}_3$, and $f(x) = x^3 - 2x$ over $\mathbb{F}_5$.

Exercises

Exercise 2.13. Count the number of solutions in $\mathbb{F}_p$ to the equation

$$\frac{x}{y} + \frac{y}{z} + \frac{z}{x} = 3.$$

Exercise 2.14. Let $a \in \mathbb{F}_p^*$. Count the number of solutions to the equation

$$(x + y + z + 1)^2 = axyz.$$

Exercise 2.15. Let $c \in \mathbb{F}_p^*$. Count the number of solutions to the equation

$$x^2 y + y^2 z + z^2 t + ct^2 x = 0.$$

Exercise 2.16. Let $p > 3$ and $a, b \in \mathbb{F}_p$. Count the number of solutions to the system of equations

$$\begin{cases} x^3 + y^3 + z^3 + t^3 = a, \\ x + y + z + t = b. \end{cases}$$

Exercise 2.17. Let $a, b, c \in \mathbb{F}_p$. Show that there exist matrices $A, B \in \mathrm{SL}_2(\mathbb{F}_p)$ satisfying the tracial conditions $\operatorname{tr} A = a$, $\operatorname{tr} B = b$, $\operatorname{tr} AB = c$.

Exercise 2.18. Let $p > 3$ and $a, b \in \mathbb{F}_p$ with $a \neq 0$. Show that the equation

$$y^2 = x^3 + ax + b$$

has at least $(p - 2)/3$ solutions.

Exercise 2.19. Consider the quartic map $f(x) = x^4 + ax^2$ on $\mathbb{F}_p$, where $a \neq 0$. Show that the number of values of f is given as follows:

$$\#f(\mathbb{F}_p) = \begin{cases} \lceil 3p/8 \rceil & \text{if } p \equiv 1, 7 \bmod 8, \\ \lceil 3p/8 \rceil - (1 - \sigma(a))/2 & \text{if } p \equiv 3 \bmod 8, \\ \lceil 3p/8 \rceil + (1 - \sigma(a))/2 & \text{if } p \equiv 5 \bmod 8. \end{cases}$$

Notes

- Theorem 2.4 is due to Jacobsthal (*Anwendungen einer Formel aus der Theorie der quadratischen Reste*, Dissertation, Berlin, 1906). It is by now a well known fact, to the point that its appearances in the mathematical literature rarely credit Jacobsthal.

- The Iverson bracket notation, $[\![\cdot]\!]$, is a propositional enhancement of Kronecker's delta notation, though much less used. A great case for Iverson's bracket is made by Knuth in *Two notes on notation* (Amer. Math. Monthly 1992). Iverson also introduced the floor and ceiling functions and their commonly used notations, $\lfloor \cdot \rfloor$ and $\lceil \cdot \rceil$. Iverson's ingenious notations are surprisingly recent–they date from the early 1960's.

- Theorem 2.10 is due to von Sterneck (*Über die Anzahl inkongruenter Werte, die eine ganze Funktion dritten Grades annimmt*, Sitzungsber. Akad. Wiss. Wien 1907). It reveals an unexpected uniformity: for any $\alpha \neq 0$, a cubic map of the form $x^3 + \alpha x$ takes on the same the number of values!

Extending Theorem 2.10 to quartic maps is difficult. Proceeding as in the proof of Theorem 2.10, one reaches the analogue of (44), now a cubic equation in y; reading off the number of distinct roots is quite complicated. Exercise 2.19, also due to von Sterneck (*op. cit.*), is the simplest example of a non-trivial quartic map for which the same strategy does work. More technical results on counting values of quartics can be found in work of Zhi-Hong Sun (*On the number of incongruent residues of $x^4 + ax^2 + bx$ modulo p*, J. Number Theory 2006).

- Example 2.7 and Exercise 2.14 are results due to Carlitz (*Certain special equations in a finite field*, Monatsh. Math. 1954; *The number of solutions of some equations in a finite field*, Portugal. Math. 1954).

- The Fricke identity says that

$$(\operatorname{tr} A)^2 + (\operatorname{tr} B)^2 + (\operatorname{tr} AB)^2 - (\operatorname{tr} A)(\operatorname{tr} B)(\operatorname{tr} AB) = \operatorname{tr}(ABA^{-1}B^{-1}) + 2,$$

whenever $A, B \in \mathrm{SL}_2(\mathbb{F}_p)$. We may interpret this identity as saying that the function $\tau : \mathrm{SL}_2(\mathbb{F}_p) \times \mathrm{SL}_2(\mathbb{F}_p) \to \mathbb{F}_p^3$, given by $\tau(A, B) = (\operatorname{tr} A, \operatorname{tr} B, \operatorname{tr} AB)$, sends a pair of matrices (A, B) to a solution of (37) for $c = \operatorname{tr}(ABA^{-1}B^{-1}) + 2$. By Exercise 2.17, the map τ is surjective.

- Example 2.12 is inspired by the notion of "anomalous prime", as introduced by Mazur (*Rational points of abelian varieties with values in towers of number fields*, Invent. Math. 1972). For rational elliptic curves, this concept boils down to the following: given an elliptic equation $y^2 = f(x)$ over $\mathbb{Q}$, where f is a cubic map with coefficients in $\mathbb{Z}$, an odd prime p is said to be *anomalous* if the reduction of f modulo p is a cubic map without repeated roots in $\mathbb{F}_p$, and

$$\sum_{x \in \mathbb{F}_p} \left(\frac{f(x)}{p} \right) \equiv -1 \bmod p.$$

In this formulation, the upshot of Example 2.12 is that a rational elliptic equation of the form $y^2 = x^3 + bx^2 + cx$, where $b, c \in \mathbb{Z}$, can only have $p = 3$ and $p = 5$ as anomalous odd primes; furthermore, $p = 3$ is an anomalous prime if and only if $b \equiv 1 \bmod 3$ and $c \equiv 2 \bmod 3$, while $p = 5$ is an anomalous prime if and only if $b \equiv 0 \bmod 5$ and $c \equiv 3 \bmod 5$.

Anomalies for cubic maps of the form $f(x) = x^3 + a$ are proposed further ahead in this text, as Exercise 3.30.

Chapter 3

Jacobsthal Sums

Introducing the Jacobsthal sums φ_n and ψ_n

A *Jacobsthal sum* is a quadratic character sum having one of the following two forms:

$$\varphi_n(a) = \sum_{x \in \mathbb{F}_p} \sigma(x^{n+1} + ax) \tag{46}$$

$$\psi_n(a) = \sum_{x \in \mathbb{F}_p} \sigma(x^n + a), \tag{47}$$

where n is a positive integer, and $a \in \mathbb{F}_p$ is a parameter. We note that there is a notational shift in the φ_n-sum: the index, n, does not match the degree, $n+1$, of the polynomial argument. This may seem slightly puzzling at first sight but, by writing

$$\varphi_n(a) = \sum_{x \in \mathbb{F}_p} \sigma(x)\sigma(x^n + a),$$

we may come to interpret the Jacobsthal sum φ_n as the twisted twin of ψ_n.

The Jacobsthal sums for the parameter $a = 0$ are outliers in several ways. To begin with, they are easily computed:

$$\varphi_n(0) = (p-1)\,[\![n \text{ is odd}]\!], \quad \psi_n(0) = (p-1)\,[\![n \text{ is even}]\!]. \tag{48}$$

But the problem of computing Jacobsthal sums for non-zero parameters is, generally speaking, a very non-trivial task. We will also see that, for fixed n, Jacobsthal sums for parameters $a \neq 0$ have magnitude on the order of $\sqrt{p}$; by way of contrast, (48) shows that Jacobsthal sums for the parameter $a = 0$ can be on the order of p.

Before we begin in earnest our investigation of the Jacobsthal sums, let us clarify that our main interest is in small values of n. For very small values of n, we recall that

$$\psi_1(a) = 0, \quad \psi_2(a) = -1, \quad \varphi_1(a) = -1$$

for each $a \in \mathbb{F}_p^*$; these reflect the identities (5), (34) and (35). Such simple and complete answers do not work, unfortunately, for other Jacobsthal sums. But the techniques encountered in the proof of Theorem 2.4 do carry through and, with the help of additional ingredients, we will eventually reach some explicit evaluations of Jacobsthal sums in low degree. To spell it out, our results will ultimately pertain to the sums

$$\varphi_2(a) = \sum_{x \in \mathbb{F}_p} \sigma(x^3 + ax), \tag{49}$$

$$\varphi_3(a) = \sum_{x \in \mathbb{F}_p} \sigma(x^4 + ax), \tag{50}$$

$$\varphi_4(a) = \sum_{x \in \mathbb{F}_p} \sigma(x^5 + ax), \tag{51}$$

as well as

$$\psi_3(a) = \sum_{x \in \mathbb{F}_p} \sigma(x^3 + a), \tag{52}$$

$$\psi_4(a) = \sum_{x \in \mathbb{F}_p} \sigma(x^4 + a). \tag{53}$$

In the following table, we collect values of Jacobsthal sums for $p = 13$. We invite the reader to jot down some potential patterns.

a	0	1	2	3	4	5	6	7	8	9	10	11	12
$\varphi_2(a)$	0	6	-4	6	-6	-4	-4	4	4	6	-6	4	-6
$\psi_3(a)$	0	-2	5	-5	7	2	-7	-7	2	7	-5	5	-2

Properties of Jacobsthal sums I

We start with a reduction principle.

Lemma 3.1. *For each $a \in \mathbb{F}_p$, we have*

$$\psi_{2n}(a) = \psi_n(a) + \varphi_n(a). \tag{54}$$

Proof. We have

$$\psi_{2n}(a) = \sum_{x \in \mathbb{F}_p} \sigma(x^{2n} + a)$$

$$= \sum_{x \in \mathbb{F}_p} \sigma(x^n + a) + \sum_{x \in \mathbb{F}_p} \sigma(x(x^n + a)) = \psi_n(a) + \varphi_n(a),$$

by using the downgrading formula (31) for $f(x) = x^n + a$. $\quad\square$

Lemma 3.2. *Let $a \in \mathbb{F}_p$, and put $d = \gcd(n, p - 1)$. Then*

$$\psi_n(a) = \psi_d(a), \qquad \varphi_n(a) = \varphi_d(a). \tag{55}$$

Furthermore,

$$\varphi_n(a) = 0 \quad \text{if } p \not\equiv 1 \bmod 2d. \tag{56}$$

Proof. As the power functions x^n and x^d have the same solution-counting maps, we have $\psi_n(a) = \psi_d(a)$, and $\psi_{2n}(a) = \psi_{2d}(a)$. By (54), we have

$$\varphi_n(a) = \psi_{2n}(a) - \psi_n(a) = \psi_{2d}(a) - \psi_d(a) = \varphi_d(a).$$

If $2d$ does not divide $p - 1$, then $\gcd(2n, p - 1) = d$. Therefore, $\psi_{2n}(a) = \psi_d(a) = \psi_n(a)$, and so $\varphi_n(a) = \psi_{2n}(a) - \psi_n(a) = 0$. $\quad\square$

Lemma 3.2 delineates the essential values for n: we may assume without loss of generality that $p \equiv 1 \bmod n$ and, as far as the φ_n-sum is concerned, we may focus in fact on the case when $p \equiv 1 \bmod 2n$.

The next result provides information modulo n, and even modulo $2n$, about the Jacobsthal sums.

Lemma 3.3. *Let $a \in \mathbb{F}_p$.*

(i) *Assume $p \equiv 1 \bmod n$. Then $\psi_n(a) \equiv \sigma(a) \bmod n$. Furthermore,*

$$\frac{\psi_n(a) - \sigma(a)}{n} \equiv \frac{p - 1}{n} - [\![-a \in (\mathbb{F}_p^*)^n]\!] \quad \bmod 2. \tag{57}$$

(ii) *Assume $p \equiv 1 \bmod 2n$. Then $\varphi_n(a) \equiv 0 \bmod n$. Furthermore,*

$$\frac{\varphi_n(a)}{n} \equiv [\![-a \in (\mathbb{F}_p^*)^n]\!] \quad \bmod 2. \tag{58}$$

Proof.   (i) We write

$$\psi_n(a) - \sigma(a) = \sum_{x \in \mathbb{F}_p^*} \sigma(x^n + a) = n \sum_{s \in (\mathbb{F}_p^*)^n} \sigma(s + a)$$

since the equation $x^n = s$ has n solutions whenever $s \in (\mathbb{F}_p^*)^n$. This follows from Theorem 1.3, keeping in mind that n divides $p - 1$. At this point, we have $\psi_n(a) - \sigma(a) \equiv 0 \bmod n$, thereby proving the first claim.

For the second claim, we write

$$\frac{\psi_n(a) - \sigma(a)}{n} = \sum_{s \in (\mathbb{F}_p^*)^n} \sigma(s + a)$$

and we find the parity of the right-hand sum. The number of terms is $(p-1)/n$. If $-a \notin (\mathbb{F}_p^*)^n$ then each term is ± 1, so the right-hand sum is $(p-1)/n \bmod 2$. If $-a \in (\mathbb{F}_p^*)^n$ then one term vanishes and all the remaining terms are ± 1, so the right-hand sum is $(p-1)/n - 1 \bmod 2$. The two cases can be combined into (57).

(ii) Again, we exploit the identity $\varphi_n(a) = \psi_{2n}(a) - \psi_n(a)$, in conjunction with part (i). As $p \equiv 1 \bmod 2n$, we have $\psi_{2n}(a) \equiv \sigma(a) \bmod 2n$; in particular, $\psi_{2n}(a) \equiv \sigma(a) \bmod n$. Also, $p \equiv 1 \bmod n$ and so $\psi_n(a) \equiv \sigma(a) \bmod n$. It follows that $\varphi_n(a) \equiv 0 \bmod n$. Moreover, we write

$$\frac{\varphi_n(a)}{n} = \frac{\psi_{2n}(a) - \sigma(a)}{n} - \frac{\psi_n(a) - \sigma(a)}{n}.$$

The first term on the right-hand side is even. For the second term, we have

$$\frac{\psi_n(a) - \sigma(a)}{n} \equiv \frac{p-1}{n} - [\![-a \in (\mathbb{F}_p^*)^n]\!] \equiv [\![-a \in (\mathbb{F}_p^*)^n]\!] \quad \bmod 2.$$

The congruence (58) follows. $\square$

Example 3.4.   Assume $p \equiv 1 \bmod n$. We have

$$\frac{\psi_n(1) - 1}{n} \equiv \frac{p-1}{n} - [\![-1 \in (\mathbb{F}_p^*)^n]\!] \quad \bmod 2.$$

As $-1 \in (\mathbb{F}_p^*)^n$ if and only if $(p-1)/n$ is even, we deduce that $(\psi_n(1)-1)/n$ is odd.

Assuming $p \equiv 1 \bmod 2n$, we have $[\![-1 \in (\mathbb{F}_p^*)^n]\!] = 1$ and so $\varphi_n(1)/n$ is odd.

Lemma 3.5 (Inversion). *Let $a \in \mathbb{F}_p^*$.*

(i) *Assume n is even. Then*

$$\psi_n(a^{-1}) = \sigma(a)\psi_n(a) + \sigma(a) - 1, \quad \varphi_n(a^{-1}) = \sigma(a)\varphi_n(a). \qquad (59)$$

(ii) *Assume n is odd. Then*

$$\psi_n(a^{-1}) = \sigma(a)\varphi_n(a) + \sigma(a). \qquad (60)$$

Proof. Using the change of variable $x := x^{-1}$ over $\mathbb{F}_p^*$, we write:

$$\psi_n(a^{-1}) - \sigma(a^{-1}) = \sum_{x \in \mathbb{F}_p^*} \sigma(x^n + a^{-1}) = \sum_{x \in \mathbb{F}_p^*} \sigma(x^{-n} + a^{-1})$$

$$= \sigma(a) \sum_{x \in \mathbb{F}_p^*} \sigma(x^n)\sigma(x^n + a).$$

When n is even, the latter sum equals $\psi_n(a) - \sigma(a)$; thus $\psi_n(a^{-1}) - \sigma(a) = \sigma(a)\psi_n(a) - 1$, proving the first identity in (59). When n is odd, the latter sum equals $\varphi_n(a)$; so $\psi_n(a^{-1}) - \sigma(a) = \sigma(a)\varphi_n(a)$, thereby proving (60).

We turn to the second identity in (59). Again, we use the change of variable $x := x^{-1}$ over $\mathbb{F}_p^*$:

$$\varphi_n(a^{-1}) = \sum_{x \in \mathbb{F}_p^*} \sigma(x^{n+1} + a^{-1}x) = \sum_{x \in \mathbb{F}_p^*} \sigma\big(x^{-(n+1)} + a^{-1}x^{-1}\big)$$

$$= \sigma(a) \sum_{x \in \mathbb{F}_p^*} \sigma(x^{n+1})\sigma(x^n + a).$$

When n is even, the latter sum equals $\varphi_n(a)$; whence $\varphi_n(a^{-1}) = \sigma(a)\varphi_n(a)$. When n is odd, the latter sum equals $\psi_n(a) - \sigma(a)$; we then run into an identity equivalent to (60). $\qquad \square$

Among properties of the Jacobsthal sums φ_n and ψ_n, the identities (54) and (60) stand out as the only ones relating the two types of Jacobsthal sums. We illustrate them, in low degree, in the following example.

Example 3.6. Let $a \in \mathbb{F}_p^*$. Taking $n = 2$ in (54), and recalling that $\psi_2(a) = -1$, we get

$$\psi_4(a) = -1 + \varphi_2(a),$$

explicitly,

$$\sum_{x \in \mathbb{F}_p} \sigma(x^4 + a) = -1 + \sum_{x \in \mathbb{F}_p} \sigma(x^3 + ax).$$

Taking $n = 3$ in (60), we obtain

$$\varphi_3(a) = -1 + \sigma(a)\psi_3(a^{-1}),$$

explicitly,

$$\sum_{x \in \mathbb{F}_p} \sigma(x^4 + ax) = -1 + \sigma(a) \sum_{x \in \mathbb{F}_p} \sigma(x^3 + a^{-1}).$$

The parity of n played a role in the previous lemma; it does again, in the next property.

Lemma 3.7 (Periodicity). *Let $a \in \mathbb{F}_p$, and $s \in \mathbb{F}_p^*$.*

(i) *If n is even, then*

$$\psi_n(s^n a) = \psi_n(a), \qquad \varphi_n(s^n a) = \sigma(s)\varphi_n(a). \tag{61}$$

(ii) *If n is odd, then*

$$\psi_n(s^n a) = \sigma(s)\psi_n(a), \qquad \varphi_n(s^n a) = \varphi_n(a). \tag{62}$$

Proof. Using the change of variable $x := sx$, we write:

$$\psi_n(s^n a) = \sum_{x \in \mathbb{F}_p} \sigma(x^n + s^n a) = \sum_{x \in \mathbb{F}_p} \sigma(s^n x^n + s^n a)$$

$$= \sigma(s)^n \sum_{x \in \mathbb{F}_p} \sigma(x^n + a) = \sigma(s)^n \psi_n(a),$$

respectively,

$$\varphi_n(s^n a) = \sum_{x \in \mathbb{F}_p} \sigma(x^{n+1} + s^n ax) = \sum_{x \in \mathbb{F}_p} \sigma(s^{n+1} x^{n+1} + s^{n+1} ax)$$

$$= \sigma(s)^{n+1} \sum_{x \in \mathbb{F}_p} \sigma(x^{n+1} + ax) = \sigma(s)^{n+1} \varphi_n(a).$$

Taking the parity of n into account, we get (61) and (62). $\qquad \square$

Corollary 3.8 (Sign change). *Assume $p \equiv 1 \bmod 2n$, and let $a \in \mathbb{F}_p$. Then*

$$\varphi_n(-a) = \begin{cases} \varphi_n(a) & \text{if } n \text{ is odd,} \\ (-1)^{(p-1)/(2n)}\varphi_n(a) & \text{if } n \text{ is even.} \end{cases} \tag{63}$$

Proof. The main point is that -1 is an n-th power: if g is a generator of the cyclic group $\mathbb{F}_p^*$, then $s = g^{(p-1)/(2n)}$ satisfies $s^n = -1$.

When n is odd, we have $\varphi_n(-a) = \varphi_n(s^n a) = \varphi_n(a)$ by (62). When n is even, we have $\varphi_n(-a) = \varphi_n(s^n a) = \sigma(s)\varphi_n(a)$ by (61); and $\sigma(s) = (-1)^{(p-1)/(2n)}$, since $\sigma(g) = -1$. $\square$

Example 3.9. When $p \equiv 1 \bmod 4$, we have

$$\varphi_2(-a) = (-1)^{(p-1)/4}\varphi_2(a). \tag{64}$$

A variation of (64) that works for all p is

$$\varphi_2(-a) = \sigma(2)\varphi_2(a). \tag{65}$$

Indeed, when $p \equiv 3 \bmod 4$ the Jacobsthal sum φ_2 is identically 0; thus (65) clearly holds. When $p \equiv 1 \bmod 4$, we have $\sigma(2) = (-1)^{(p^2-1)/8} = (-1)^{(p-1)/4}$; in this case (65) replicates (64).

Properties of Jacobsthal sums II

Lemma 3.10 (First moments). *We have*

$$\sum_{a \in \mathbb{F}_p} \psi_n(a) = 0, \qquad \sum_{a \in \mathbb{F}_p} \varphi_n(a) = 0. \tag{66}$$

In statistical terms, (66) says that both Jacobsthal sums have expected value 0.

Proof. By interchanging the order of summation, we have

$$\sum_{a \in \mathbb{F}_p} \psi_n(a) = \sum_{a \in \mathbb{F}_p} \sum_{x \in \mathbb{F}_p} \sigma(x^n + a) = \sum_{x \in \mathbb{F}_p} \sum_{a \in \mathbb{F}_p} \sigma(x^n + a) = 0,$$

$$\sum_{a \in \mathbb{F}_p} \varphi_n(a) = \sum_{a \in \mathbb{F}_p} \sum_{x \in \mathbb{F}_p^*} \sigma(x)\sigma(x^n + a) = \sum_{x \in \mathbb{F}_p^*} \sigma(x) \sum_{a \in \mathbb{F}_p} \sigma(x^n + a) = 0,$$

as each inner sum vanishes. $\square$

Lemma 3.11 (Second moments). *The following hold:*

(i) *if $p \equiv 1 \bmod n$ then*

$$\sum_{a \in \mathbb{F}_p} \psi_n(a)^2 = (n-1)(p^2 - p); \tag{67}$$

(ii) *if $p \equiv 1 \bmod 2n$ then*

$$\sum_{a \in \mathbb{F}_p} \varphi_n(a)^2 = n(p^2 - p). \tag{68}$$

Proof. (i) We have

$$\sum_{a \in \mathbb{F}_p} \psi_n(a)^2 = \sum_{a \in \mathbb{F}_p} \left(\sum_{x \in \mathbb{F}_p} \sigma(x^n + a) \right) \left(\sum_{y \in \mathbb{F}_p} \sigma(y^n + a) \right)$$

$$= \sum_{x,y \in \mathbb{F}_p} \sum_{a \in \mathbb{F}_p} \sigma(a + x^n)\sigma(a + y^n).$$

The inner sum equals $p[\![x^n = y^n]\!] - 1$, by (36). Thus,

$$\sum_{a \in \mathbb{F}_p} \psi_n(a)^2 = \sum_{x,y \in \mathbb{F}_p} \left(p[\![x^n = y^n]\!] - 1 \right)$$

$$= p \cdot \#\{(x,y) : x^n = y^n\} - p^2.$$

With the exception of $(0,0)$, the solutions to $x^n = y^n$ are parameterized as (x, xz), where $x \in \mathbb{F}_p^*$, and $z^n = 1$. There are exactly n nth roots of unity in $\mathbb{F}_p$, so $\#\{(x,y) : x^n = y^n\} = 1 + n(p-1)$. Formula (67) follows.

(ii) The argument is similar, though slightly more elaborate. We have

$$\sum_{a \in \mathbb{F}_p} \varphi_n(a)^2 = \sum_{a \in \mathbb{F}_p} \left(\sum_{x \in \mathbb{F}_p} \sigma(x^{n+1} + ax) \right) \left(\sum_{y \in \mathbb{F}_p} \sigma(y^{n+1} + ay) \right)$$

$$= \sum_{x,y \in \mathbb{F}_p} \sigma(xy) \sum_{a \in \mathbb{F}_p} \sigma(a + x^n)\sigma(a + y^n)$$

$$= \sum_{x,y \in \mathbb{F}_p} \sigma(xy)\left(p[\![x^n = y^n]\!] - 1 \right) = p \sum_{\substack{x,y \in \mathbb{F}_p \\ x^n = y^n}} \sigma(xy),$$

as $\sum_{x,y\in\mathbb{F}_p}\sigma(xy) = \left(\sum_{x\in\mathbb{F}_p}\sigma(x)\right)\left(\sum_{y\in\mathbb{F}_p}\sigma(y)\right) = 0$. The latter sum can be evaluated as follows:

$$\sum_{\substack{x,y\in\mathbb{F}_p \\ x^n=y^n}}\sigma(xy) = \sum_{x\in\mathbb{F}_p^*}\sum_{\substack{z\in\mathbb{F}_p \\ z^n=1}}\sigma(x(xz)) = (p-1)\sum_{\substack{z\in\mathbb{F}_p \\ z^n=1}}\sigma(z).$$

Now, all the n nth roots of unity are squares since $(p-1)/n$ is even. So, the latter sum equals n, thereby yielding (68). $\qquad\square$

Corollary 3.12. *For each $a \in \mathbb{F}_p^*$, we have*

$$|\psi_n(a)|, \ |\varphi_n(a)| \leq \gcd(n, p-1)\sqrt{p}. \tag{69}$$

Proof. Recall, from (55), that replacing n by $\gcd(n, p-1)$ leaves both Jacobsthal sums unchanged. So it suffices to prove that

$$|\psi_n(a)|, \ |\varphi_n(a)| \leq n\sqrt{p},$$

whenever $p \equiv 1 \bmod n$. In what concerns the φ_n-sum, we may furthermore assume that $p \equiv 1 \bmod 2n$; for otherwise $\varphi_n(a) = 0$, by (56), so the bound holds trivially.

By the periodicity of the Jacobsthal sums, each term in the identities (67) and (68) appears with a certain multiplicity. Indeed, fix $a \in \mathbb{F}_p^*$; from (61) and (62), we deduce that $\psi_n(ca)^2 = \psi_n(a)^2$ and $\varphi_n(ca)^2 = \varphi_n(a)^2$, whenever $c \in (\mathbb{F}_p^*)^n$. Note that $\#(\mathbb{F}_p^*)^n = (p-1)/n$. Thus, the term $\psi_n(a)^2$ appears $(p-1)/n$ times on the left-hand side of (67) and, similarly, $\varphi_n(a)^2$ appears $(p-1)/n$ times on the left-hand side of (68). We deduce that

$$\frac{p-1}{n}\,\psi_n(a)^2 \leq (n-1)(p^2 - p)$$

implying that $\psi_n(a)^2 < n^2 p$, or $|\psi_n(a)| < n\sqrt{p}$. Similarly,

$$\frac{p-1}{n}\,\varphi_n(a)^2 \leq n(p^2 - p),$$

so $\varphi_n(a)^2 \leq n^2 p$, that is, $|\varphi_n(a)| \leq n\sqrt{p}$. $\qquad\square$

The next fact, also a consequence of Lemma 3.11, is a lower bound counterpart to (69). Its main upshot is that upper bounds for Jacobsthal sums on the order of $\sqrt{p}$ are best possible.

Corollary 3.13. *The following hold:*

(i) *if $p \equiv 1 \bmod n$, then $|\psi_n(a)| > \sqrt{(n-2)p}$ for some $a \in \mathbb{F}_p^*$;*

(ii) *if $p \equiv 1 \bmod 2n$, then $|\varphi_n(a)| > \sqrt{(n-1)p}$ for some $a \in \mathbb{F}_p^*$.*

Proof. (i) Let $a^* \in \mathbb{F}_p^*$ with $|\psi_n(a^*)| = \max_{a \in \mathbb{F}_p^*} |\psi_n(a)|$. By (67),

$$(p-1)\psi_n(a^*)^2 \geq \sum_{a \in \mathbb{F}_p^*} \psi_n(a)^2 = (n-1)(p^2 - p) - \psi_n(0)^2.$$

As $\psi_n(0)$ equals 0 or $p - 1$, we deduce that

$$(p-1)\psi_n(a^*)^2 \geq (n-1)(p^2 - p) - (p-1)^2 = (p-1)\big((n-2)p + 1\big).$$

It follows that $|\psi_n(a^*)| > \sqrt{(n-2)p}$.

(ii) The argument, virtually identical to (i), is left to the reader. $\qquad\square$

Representations of p as sums of squares

The previous two sections offer a tangle of relations pertaining to Jacobsthal sums. In this section, we wish to take full stock of their consequences. For the sake of clarity we focus on the φ_n sums. We assume that $p \equiv 1 \bmod 2n$ in what follows.

A key role is played by the second moment formula (68). We aim to use the periodicity of the Jacobsthal sums φ_n in order to fold it into much fewer terms. A related idea was already exploited in the proof of Corollary 3.12.

Let g be a generator of the multiplicative group $\mathbb{F}_p^*$. We put

$$w_k = \varphi_n(g^k). \tag{70}$$

Here, k is, *a priori*, an integer, though the natural range is $k = 0, \ldots, p - 2$ thanks to the obvious $(p-1)$-periodicity of the w-sequence. Notably,

$$w_0 = \varphi_n(1). \tag{71}$$

The following lemma translates, in the language of the w-sequence, all the relations for the Jacobsthal sums φ_n that we have previously established.

Lemma 3.14. *The integers w_k enjoy the following properties:*

$$\sum_{k=0}^{p-2} w_k = \begin{cases} -(p-1) & \text{if } n \text{ is odd,} \\ 0 & \text{if } n \text{ is even,} \end{cases} \tag{72}$$

$$\sum_{k=0}^{p-2} w_k^2 = \begin{cases} n(p^2 - p) - (p-1)^2 & \text{if } n \text{ is odd,} \\ n(p^2 - p) & \text{if } n \text{ is even,} \end{cases} \tag{73}$$

$$w_k \equiv 0 \bmod n, \ w_k/n \equiv 1 \bmod 2 \ \textit{if and only if } n \mid k, \tag{74}$$

$$w_{n+k} = \begin{cases} w_k & \text{if } n \text{ is odd,} \\ -w_k & \text{if } n \text{ is even,} \end{cases} \tag{75}$$

$$w_{-k} = (-1)^k w_k \qquad \text{if } n \text{ is even.} \tag{76}$$

Proof. The first moment identity (66) and the second moment identity (68) give

$$-\varphi_n(0) = \sum_{a \in \mathbb{F}_p^*} \varphi_n(a) = \sum_{k=0}^{p-2} \varphi_n(g^k) = \sum_{k=0}^{p-2} w_k,$$

$$n(p^2 - p) - \varphi_n(0)^2 = \sum_{a \in \mathbb{F}_p^*} \varphi_n(a)^2 = \sum_{k=0}^{p-2} \varphi_n(g^k)^2 = \sum_{k=0}^{p-2} w_k^2.$$

We recall from (48) that $\varphi_n(0) = p - 1$ if n is odd, respectively, $\varphi_n(0) = 0$ if n is even. The identities (72) and (73) follow.

For (74), we use part (ii) of Lemma 3.3. We have $w_k \equiv 0 \bmod n$, and $w_k/n \equiv [\![-g^k \in (\mathbb{F}_p^*)^n]\!] \bmod 2$. Now, $-g^k \in (\mathbb{F}_p^*)^n$ if and only if $(-g^k)^{(p-1)/n} = 1$. As $(p-1)/n$ is even, this amounts to $g^{k(p-1)/n} = 1$; this, in turn, holds if and only if n divides k, since g has order $p - 1$.

For the remaining identities, we use the observation that $\sigma(g) = -1$. To get (75), we rely on the periodicity Lemma 3.7. When n is odd, we have

$$w_{n+k} = \varphi_n(g^n g^k) = \varphi_n(g^k) = w_k.$$

When n is even, we have

$$w_{n+k} = \varphi_n(g^n g^k) = \sigma(g)\varphi_n(g^k) = -w_k.$$

Lastly, we deduce (76) from (59) as follows:

$$w_{-k} = \varphi_n(g^{-k}) = \sigma(g^k)\varphi_n(g^k) = (-1)^k w_k.$$

This completes the proof. $\qquad\qquad\qquad\qquad\qquad\qquad\qquad\qquad\qquad\square$

The twisted periodicity (75) implies that the essential range for the w-sequence is actually $k = 0, \ldots, n - 1$. A notable consequence is that the sum-of-squares identity (73) can be compressed from $p - 1$ terms to n terms. But the lemma also highlights a parity issue. We therefore delineate the case when n is odd from the case when n is even. In the latter case, the additional relation (76) catches the eye; this apparent imbalance is evened out, however, since the vanishing relation (72) is made redundant by (75).

Corollary 3.15. *Assume $p \equiv 1 \bmod 2n$. Then the integers $w_0, \ldots, w_{n-1}$ are multiples of n, and w_k/n is odd for $k = 0$ only. Furthermore,*

(i) *if n is even, then*

$$\sum_{k=0}^{n-1} w_k^2 = n^2 p, \qquad w_{n-k} = (-1)^{k+1} w_k;$$

(ii) *if n is odd, then*

$$\sum_{k=0}^{n-1} w_k^2 = n^2 p - n(p-1), \qquad \sum_{k=0}^{n-1} w_k = -n.$$

Proof. The first statement reflects (74) in the range $k = 0, \ldots, n - 1$.

(i) If n is even, then $w_{n+k} = -w_k$ for each integer k. Hence, $w_{n-k} = -w_{-k} = -(-1)^k w_k$ for each $k = 0, \ldots, n - 1$, by (76). Also, $w_{n+k}^2 = w_k^2$ for each integer k whence

$$np(p-1) = \sum_{k=0}^{p-2} w_k^2 = \frac{p-1}{n} \sum_{k=0}^{n-1} w_k^2.$$

The compressed sum-of-squares identity follows.

(ii) If n is odd, then $w_{n+k} = w_k$ for each integer k. Therefore,

$$np(p-1) - (p-1)^2 = \sum_{k=0}^{p-2} w_k^2 = \frac{p-1}{n} \sum_{k=0}^{n-1} w_k^2,$$

and

$$-(p-1) = \sum_{k=0}^{p-2} w_k = \frac{p-1}{n} \sum_{k=0}^{n-1} w_k.$$

The compressed sums are thereby evaluated. $\qquad\square$

The sum-of-squares identities with the fewest terms arise for $n = 2, 3, 4$. We spell out the outcome of the previous lemma for these small values of n.

Theorem 3.16. *Let g be a generator of the multiplicative group $\mathbb{F}_p^*$.*

(i) *Assume that $p \equiv 1 \bmod 4$. Then $A^2 + B^2 = p$ for the integers A and B defined by*

$$A = \frac{\varphi_2(1)}{2}, \qquad B = \frac{\varphi_2(g)}{2}.$$

Furthermore, A is odd and B is even.

(ii) *Assume that $p \equiv 1 \bmod 6$. Then $A^2 + 3B^2 = p$ for the integers A and B defined by*

$$A = \frac{1 + \varphi_3(1)}{2}, \qquad B = \frac{\varphi_3(g) - \varphi_3(g^2)}{6}.$$

Furthermore, $A \equiv 2 \bmod 3$.

(iii) *Assume that $p \equiv 1 \bmod 8$. Then $A^2 + 2B^2 = p$ for the integers A and B defined by*

$$A = \frac{\varphi_4(1)}{4}, \qquad B = \frac{\varphi_4(g)}{4}.$$

Furthermore, A is odd and B is even.

Proof. We refer to Corollary 3.15, whose notations we keep.

(i) For $n = 2$, the integers w_0 and w_1 are even, and satisfy $w_0^2 + w_1^2 = 4p$. For $A = w_0/2$ and $B = w_1/2$, we get

$$A^2 + B^2 = p.$$

Moreover, $A = w_0/2$ is odd and $B = w_1/2$ is even.

(ii) For $n = 3$, the integers w_0, w_1, w_2 are multiples of 3. Additionally, $w_0/3$ is odd, while $w_1/3$ and $w_2/3$ are even; whence w_0 is odd, and w_1, w_2 are even. Therefore, $w_0 \equiv 3 \bmod 6$, and $w_1, w_2 \equiv 0 \bmod 6$. We also know that $w_0^2 + w_1^2 + w_2^2 = 6p + 3$, and $w_0 + w_1 + w_2 = -3$. In terms of $v_k = 1 + w_k$, we have

$$v_0^2 + v_1^2 + v_2^2 = 6p, \qquad v_0 + v_1 + v_2 = 0.$$

Therefore,

$$12p = 2v_0^2 + 2v_1^2 + 2v_2^2$$
$$= 2v_0^2 + (v_1 + v_2)^2 + (v_1 - v_2)^2 = 3v_0^2 + (v_1 - v_2)^2.$$

Now, $A = v_0/2$ and $B = (v_1 - v_2)/6$ so

$$A^2 + 3B^2 = p.$$

We note that B is an integer, since $v_1 - v_2 \equiv 0 \bmod 6$, and A is integer satisfying $A \equiv 2 \bmod 3$, since $v_0 \equiv 4 \bmod 6$.

(iii) For $n = 4$, the integers w_0, w_1, w_2, w_3 are multiples of 4, and they satisfy $w_3 = w_1$ and $w_2 = -w_2$. Thus, $w_2 = 0$. They also satisfy $w_0^2 + w_1^2 + w_2^2 + w_3^2 = 16p$, which reduces to $w_0^2 + 2w_1^2 = 16p$. For $A = w_0/4$ and $B = w_1/4$, we obtain

$$A^2 + 2B^2 = p.$$

Moreover, $A = w_0/4$ is odd and $B = w_1/4$ is even. $\qquad\square$

We have thus obtained explicit representations of p as sums of squares. Theorem 3.16 covers three of the four modular cases appearing in Theorem 1.26; the only case it misses is $p \equiv 3 \bmod 8$.

Three evaluations of Jacobsthal sums

One way to interpret Theorem 3.16 is that it achieves representations of p as a sum of (repeated) integral squares, by means of the Jacobsthal φ_n-sums. But we can also turn this interpretation on its head! Namely, we can use representations of p as sums of squares in order to evaluate Jacobsthal sums via Theorem 3.16.

From the outset, we can foresee some sign ambiguities. Indeed, the integers A and B which feature in a representation $p = A^2 + dB^2$ are only defined up to sign. We introduce the following normalization of the A-part:

$A_2(p)$: if $p \equiv 1 \bmod 4$, we let $A = A_2(p)$ be the integer defined by writing
$\qquad p = A^2 + B^2$ with $A \equiv -1 \bmod 4$,

$A_3(p)$: if $p \equiv 1 \bmod 6$, we let $A = A_3(p)$ be the integer defined by writing $p = A^2 + 3B^2$ with $A \equiv -1 \bmod 3$,

$A_4(p)$: if $p \equiv 1 \bmod 8$, we let $A = A_4(p)$ be the integer defined by writing $p = A^2 + 2B^2$ with $A \equiv -1 \bmod 4$.

In each case, the integer $A_k(p)$ is well defined. Firstly, Lemma 1.27 implies that A is determined up to sign by the sum-of-squares representation of p; the exception is the representation $p = A^2 + B^2$ where the parity of A must be specified in order to distinguish it from B. Secondly, the sign of A is resolved by the additional modular condition.

Below we tabulate the A_k-parts for the first five primes satisfying $p \equiv 1 \bmod 2k$, for $k = 2, 3, 4$.

$p \equiv 1 \bmod 4$	$A_2(p)$	$p \equiv 1 \bmod 6$	$A_3(p)$	$p \equiv 1 \bmod 8$	$A_4(p)$
5	-1	7	2	17	3
13	3	13	-1	41	3
17	-1	19	-4	73	-1
29	-5	31	2	89	-9
37	-1	37	5	97	-5

With this viewpoint, we may interpret the representations of Theorem 3.16 as follows.

Theorem 3.17 (Evaluations up to sign).

(i) *Let $p \equiv 1 \bmod 4$, and write $p = A^2 + B^2$, where $A = A_2(p)$. Then*

$$\varphi_2(a) = \begin{cases} \pm 2A_2(p) & \text{if } a \in (\mathbb{F}_p^*)^2, \\ \pm 2B & \text{if } a \in \mathbb{F}_p^* \backslash (\mathbb{F}_p^*)^2. \end{cases}$$

(ii) *Let $p \equiv 1 \bmod 6$, and write $p = A^2 + 3B^2$, where $A = A_3(p)$. Then*

$$\varphi_3(a) = \begin{cases} 2A_3(p) - 1 & \text{if } a \in (\mathbb{F}_p^*)^3, \\ -A_3(p) - 1 \pm 3B & \text{if } a \in \mathbb{F}_p^* \backslash (\mathbb{F}_p^*)^3. \end{cases}$$

(iii) *Let $p \equiv 1 \bmod 8$, and write $p = A^2 + 2B^2$, where $A = A_4(p)$. Then*

$$
\varphi_4(a) = \begin{cases} \pm 4A_4(p) & \text{if } a \in (\mathbb{F}_p^*)^4, \\ 0 & \text{if } a \in (\mathbb{F}_p^*)^2 \backslash (\mathbb{F}_p^*)^4, \\ \pm 4B & \text{if } a \in \mathbb{F}_p^* \backslash (\mathbb{F}_p^*)^2. \end{cases}
$$

Proof. Throughout, we refer to Theorem 3.16 and its proof, whose notations we keep.

(i) Assume that $p \equiv 1 \bmod 4$. By part (i) of Theorem 3.16, we have

$$
\frac{\varphi_2(1)}{2} = \pm A_2(p), \qquad \frac{\varphi_2(g)}{2} = \pm B.
$$

Therefore, $\varphi_2(1) = \pm 2A_2(p)$ and $\varphi_2(g) = \pm 2B$. Now $\varphi_2(a) = \pm\varphi_2(1)$ when $a \in \mathbb{F}_p^*$ is a square, respectively, $\varphi_2(a) = \pm\varphi_2(g)$ when $a \in \mathbb{F}_p^*$ is not a square.

(ii) Assume that $p \equiv 1 \bmod 6$. By part (ii) of Theorem 3.16, we have

$$
\frac{1 + \varphi_3(1)}{2} = A_3(p), \qquad \frac{\varphi_3(g) - \varphi_3(g^2)}{6} = \pm B.
$$

Thus, $\varphi_3(1) = 2A_3(p) - 1$. Let us recall that $\varphi_3(1) + \varphi_3(g) + \varphi_3(g^2) = -3$. The two relations, $\varphi_3(g) + \varphi_3(g^2) = -2A_3(p) - 2$ and $\varphi_3(g) - \varphi_3(g^2) = \pm 6B$, imply that $\varphi_3(g) = -A_3(p) - 1 \pm 3B$ and $\varphi_3(g^2) = -A_3(p) - 1 \mp 3B$.

But $\varphi_3(1)$, $\varphi_3(g)$, and $\varphi_3(g^2)$ are the only three possible values for $\varphi_3(a)$, as a ranges over $\mathbb{F}_p^*$. Specifically, $\varphi_3(a) = \varphi_3(1)$ when $a \in \mathbb{F}_p^*$ is a cube, respectively $\varphi_3(a) = \varphi_3(g)$ or $\varphi_3(a) = \varphi_3(g^2)$ when $a \in \mathbb{F}_p^*$ is not a cube.

(iii) Assume that $p \equiv 1 \bmod 8$. By part (iii) of Theorem 3.16, we have

$$
\frac{\varphi_4(1)}{4} = \pm A_4(p), \qquad \frac{\varphi_4(g)}{4} = \pm B.
$$

Therefore, $\varphi_4(1) = \pm 4A_4(p)$ and $\varphi_4(g) = \pm 4B$. In the proof of Theorem 3.16 (iii) we have also seen that $\varphi_4(g^2) = 0$, and that $\varphi_4(g^3) = \varphi_4(g)$.

Now $\varphi_4(a) = \pm\varphi_4(1)$ whenever $a \in \mathbb{F}_p^*$ is a fourth power; $\varphi_4(a) = \pm\varphi_4(g^2)$ whenever $a \in \mathbb{F}_p^*$ is a square but not a fourth power; $\varphi_4(a) = \pm\varphi_4(g) = \pm\varphi_4(g^3)$ whenever $a \in \mathbb{F}_p^*$ is not a square. $\qquad \square$

Nearly all evaluations in the previous theorem are up to sign. One bright spot is the exact evaluation for $\varphi_3(1)$. Our next step is to remove the sign ambiguity for $\varphi_2(1)$ and $\varphi_4(1)$. We prove the following.

Theorem 3.18. *The following hold.*

(i) *If $p \equiv 1 \bmod 4$, then $\varphi_2(1) = 2A_2(p)$.*
(ii) *If $p \equiv 1 \bmod 6$, then $\varphi_3(1) = 2A_3(p) - 1$.*
(iii) *If $p \equiv 1 \bmod 8$, then $\varphi_4(1) = (-1)^{(p-1)/8} \cdot 4A_4(p)$.*

Actually, (ii) just spells out what we already know from Theorem 3.17 (ii); we include it for the sake of completeness. To determine the signs in the other two vexing cases, (i) and (iii), we need to know the values of $\varphi_2(1)/2$ and $\varphi_4(1)/4$ modulo 4. The key insight comes from the following fact.

Lemma 3.19. *Assume $p \equiv 1 \bmod 4$. Then $\varphi_2(-1) \equiv p - 3 \bmod 16$.*

Proof. Let

$$S = \sum_{x \in \mathbb{F}_p} \left(1 - \sigma(x - 1)\right)\left(1 - \sigma(x)\right)\left(1 - \sigma(x + 1)\right).$$

We interpret the sum S by considering the subset

$$NNN = \{x \in \mathbb{F}_p : \sigma(x - 1) = \sigma(x) = \sigma(x + 1) = -1\}.$$

The somewhat unusual notation is meant to suggest that NNN captures runs of three consecutive non-squares in $\mathbb{F}_p$. If $x \in NNN$, then the corresponding summand in S equals 8. If $x = 0, \pm 1$, then the corresponding summand of S equals 0; the same is clearly true for all the remaining values of x in $\mathbb{F}_p$. The upshot is that

$$S = 8 \cdot \# NNN.$$

Observe, furthermore, that NNN is a symmetric subset of $\mathbb{F}_p^*$: if $x \in NNN$ then $-x \in NNN$ as well, since $\sigma(-1) = 1$. Consequently, the size of NNN is even, which implies that

$$S \equiv 0 \bmod 16. \tag{77}$$

On the other hand, we can compute S as follows. We multiply out each summand and we break S into several sums. Besides $\sum_{x \in \mathbb{F}_p} 1 = p$, there are several quadratic character sums that appear:

$$\sum_{x \in \mathbb{F}_p} \sigma\big(x(x - 1)(x + 1)\big) = \sum_{x \in \mathbb{F}_p} \sigma(x^3 - x) = \varphi_2(-1),$$

$$\sum_{x \in \mathbb{F}_p} \sigma(x) = \sum_{x \in \mathbb{F}_p} \sigma(x-1) = \sum_{x \in \mathbb{F}_p} \sigma(x+1) = 0,$$

$$\sum_{x \in \mathbb{F}_p} \sigma\big(x(x-1)\big) = \sum_{x \in \mathbb{F}_p} \sigma\big(x(x+1)\big) = \sum_{x \in \mathbb{F}_p} \sigma\big((x-1)(x+1)\big) = -1.$$

So we have, simply, that

$$S = p - 3 - \varphi_2(-1). \tag{78}$$

By combining (77) and (78), we infer that $\varphi_2(-1) \equiv p - 3 \bmod 16$. $\qquad\square$

Theorem 3.18 will follow from the next lemma, and the sign normalization for $A_2(p)$ and $A_4(p)$.

Lemma 3.20. *The following hold.*

(i) *Assume $p \equiv 1 \bmod 4$. Then $\varphi_2(1)/2 \equiv -1 \bmod 4$.*
(ii) *Assume $p \equiv 1 \bmod 8$. Then $\varphi_4(1)/4 \equiv -(-1)^{(p-1)/8} \bmod 4$.*

Proof. (i) By (64), we have $\varphi_2(1) = (-1)^{(p-1)/4}\varphi_2(-1)$. Lemma 3.19 yields

$$\varphi_2(1) \equiv (-1)^{(p-1)/4}(p-3) \bmod 16.$$

Now, $p \equiv 1, 5 \bmod 8$; in either case, the right-hand side is $-2 \bmod 8$. Therefore, $\varphi_2(1) \equiv -2 \bmod 8$, that is to say, $\varphi_2(1)/2 \equiv -1 \bmod 4$.

(ii) By Lemma 3.1, $\varphi_4(1) = \psi_8(1) - \psi_4(1)$ and $\psi_4(1) = -1 + \varphi_2(1)$. Hence,

$$\varphi_4(1) = \psi_8(1) + 1 - \varphi_2(1),$$

which we view modulo 16. We already know that $\varphi_2(1) \equiv p - 3 \bmod 16$. From Example 3.4, we also know that $(\psi_8(1)-1)/8$ is odd, whence $\psi_8(1) \equiv 9 \bmod 16$. Therefore,

$$\varphi_4(1) \equiv 9 + 1 - (p-3) \equiv -(p+3) \bmod 16.$$

But $p + 3 \equiv 4(-1)^{(p-1)/8} \bmod 16$, as it is easily checked by considering the two possible cases $p \equiv 1, 9 \bmod 16$. Hence, $\varphi_4(1) \equiv -4(-1)^{(p-1)/8} \bmod 16$, that is, $\varphi_4(1)/4 \equiv -(-1)^{(p-1)/8} \bmod 4$. $\qquad\square$

Very concretely, the formulas of Theorem 3.18 evaluate the following sums of Legendre symbols:

$$\sum_{x=0}^{p-1} \left(\frac{x^3 + x}{p}\right) = \begin{cases} 2A_2(p) & \text{if } p \equiv 1 \bmod 4, \\ 0 & \text{if } p \not\equiv 1 \bmod 4; \end{cases} \tag{79}$$

$$\sum_{x=0}^{p-1} \left(\frac{x^4 + x}{p}\right) = \begin{cases} 2A_3(p) - 1 & \text{if } p \equiv 1 \bmod 6, \\ -1 & \text{if } p \not\equiv 1 \bmod 6; \end{cases} \tag{80}$$

$$\sum_{x=0}^{p-1} \left(\frac{x^5 + x}{p}\right) = \begin{cases} (-1)^{(p-1)/8} \cdot 4A_4(p) & \text{if } p \equiv 1 \bmod 8, \\ 0 & \text{if } p \not\equiv 1 \bmod 8. \end{cases} \tag{81}$$

Additionally, we have $\psi_3(1) = 1 + \varphi_3(1)$ and $\psi_4(1) = -1 + \varphi_2(1)$, whence two more evaluations:

$$\sum_{x=0}^{p-1} \left(\frac{x^3 + 1}{p}\right) = \begin{cases} 2A_3(p) & \text{if } p \equiv 1 \bmod 6, \\ 0 & \text{if } p \not\equiv 1 \bmod 6; \end{cases} \tag{82}$$

$$\sum_{x=0}^{p-1} \left(\frac{x^4 + 1}{p}\right) = \begin{cases} 2A_2(p) - 1 & \text{if } p \equiv 1 \bmod 4, \\ -1 & \text{if } p \not\equiv 1 \bmod 4. \end{cases} \tag{83}$$

Building on these, one can sometimes compute other sums of Legendre symbols, as in the following example.

Example 3.21. Let $p \equiv 1 \bmod 8$. Consider the sum

$$\sum_{x=0}^{p-1} \left(\frac{x^3 + 2x}{p}\right) = \varphi_2(2).$$

As $p \equiv 1 \bmod 8$, we know that 2 is a square in $\mathbb{F}_p$; say $2 = j^2$. Therefore, $\varphi_2(2) = \varphi_2(j^2) = \sigma(j)\varphi_2(1)$. Here, $\varphi_2(1) = 2A_2(p)$. Also, $\sigma(j)$ can be made explicit: by the quartic supplementary law (25), we have

$$\sigma(j) = j^{(p-1)/2} = 2^{(p-1)/4} = (-1)^{(A_4(p)^2 - 1)/8} = (-1)^{(A_4(p)+1)/4}.$$

The last step owes to the fact that $A_4(p) \equiv -1 \bmod 4$. We conclude that

$$\sum_{x=0}^{p-1} \left(\frac{x^3 + 2x}{p}\right) = (-1)^{(A_4(p)+1)/4} \cdot 2A_2(p).$$

Three binomial coefficients mod p

A familiar fact concerning binomial coefficients mod p is that

$$\binom{p}{k} \equiv 0 \bmod p$$

for $k = 1, \ldots, p - 1$. We may visualize this fact in the reduction modulo p of Pascal's triangle: the pth row consists of 0's, except for the endpoints which are, as always, 1's. Let us consider the first p rows of Pascal's triangle mod p, from the 0th to the $(p - 1)$st; this will be referred to as *Pascal's p-triangle*. Then, in the infinite Pascal's triangle mod p, Pascal's p-triangle will repeat itself in a pattern that brings to mind the Sierpiński gasket. Below, we tabulate Pascal's p-triangle for $p = 13$.

```
                        1
                      1   1
                    1   2   1
                  1   3   3   1
                1   4   6   4   1
              1   5  10  10   5   1
            1   6   2   7   2   6   1
          1   7   8   9   9   8   7   1
        1   8   2   4   5   4   2   8   1
      1   9  10   6   9   9   6  10   9   1
    1  10   6   3   2   5   2   3   6  10   1
  1  11   3   9   5   7   7   5   9   3  11   1
1  12   1  12   1  12   1  12   1  12   1  12   1
```

The mod p computation of binomial coefficients can be reduced to entries in Pascal's p-triangle, by the *Lucas rule*: if n and k are positive integers, then

$$\binom{n}{k} \equiv \binom{n_1}{k_1}\binom{n_0}{k_0} \bmod p,$$

where $n = n_1 p + n_0$ and $k = k_1 p + k_0$ are divisions by p with remainder.

But the entries of Pascal's p-triangle are hard to understand in general–though one immediate observation is that they are all non-zero. In the next result, we give interesting formulas for certain entries lying in the middle row of Pascal's p-triangle.

Theorem 3.22. *The following hold.*

(i) *If $p \equiv 1 \bmod 4$, then*

$$\binom{\frac{1}{2}(p-1)}{\frac{1}{4}(p-1)} \equiv -2A_2(p) \bmod p. \tag{84}$$

(ii) *If $p \equiv 1 \bmod 6$, then*

$$\binom{\frac{1}{2}(p-1)}{\frac{1}{6}(p-1)} \equiv -2A_3(p) \bmod p. \tag{85}$$

(iii) *If $p \equiv 1 \bmod 8$, then*

$$\binom{\frac{1}{2}(p-1)}{\frac{1}{8}(p-1)} \equiv -(-1)^{(p-1)/8} \cdot 2A_4(p) \bmod p. \tag{86}$$

If these congruences mod p would be excised from this text, and stated on their own, it would be very hard to guess that they are obtained by means of Jacobsthal sums!

The link is provided by the following modular identity.

Lemma 3.23. *Assume $p \equiv 1 \bmod 2n$. Then*

$$\varphi_n(1) \equiv - \sum_{\substack{r=1 \\ r \text{ odd}}}^{n} \binom{\frac{1}{2}(p-1)}{\frac{r}{2n}(p-1)} \bmod p. \tag{87}$$

Proof. We have

$$\varphi_n(1) = \sum_{x \in \mathbb{F}_p} \sigma(x^{n+1} + x) \equiv \sum_{x \in \mathbb{F}_p} (x^{n+1} + x)^{(p-1)/2} \bmod p$$

thanks to Euler's identity. We evaluate the right-hand sum by using Corollary 1.2. Consider the expansion

$$(x^{n+1} + x)^{(p-1)/2} = x^{(p-1)/2}(x^n + 1)^{(p-1)/2}$$

$$= \sum_{k=0}^{(p-1)/2} \binom{\frac{1}{2}(p-1)}{k} x^{(p-1)/2 + kn}.$$

The relevant indices are those k for which $(p-1)/2 + kn$ is a multiple of $p-1$; that is to say, kn is an odd multiple of $(p-1)/2$. This means that

k is of the form $r(p-1)/(2n)$ for an odd integer $1 \le r \le n$. Thus, in $\mathbb{F}_p$, we have

$$\sum_{x \in \mathbb{F}_p} (x^{n+1} + x)^{(p-1)/2} = - \sum_{\substack{r=1 \\ r \text{ odd}}}^{n} \binom{\frac{1}{2}(p-1)}{\frac{r}{2n}(p-1)}.$$

The desired congruence (87) follows. $\qquad\square$

Proof of Theorem 3.22. We apply Lemma 3.23 for $n = 2, 3, 4$. For these small values of n we have, by Theorem 3.18, an evaluation for $\varphi_n(1)$. The other advantageous feature is that the right-hand side of (87) will contain very few terms.

(i) If $p \equiv 1 \bmod 4$, then for $n = 2$, we get

$$2A_2(p) = \varphi_2(1) \equiv - \binom{\frac{1}{2}(p-1)}{\frac{1}{4}(p-1)} \bmod p.$$

(ii) If $p \equiv 1 \bmod 6$, then for $n = 3$, we get

$$2A_3(p) - 1 = \varphi_3(1) \equiv -1 - \binom{\frac{1}{2}(p-1)}{\frac{1}{6}(p-1)} \bmod p.$$

(iii) If $p \equiv 1 \bmod 8$, then for $n = 4$, we get

$$(-1)^{(p-1)/8} \cdot 4A_4(p) = \varphi_4(1) \equiv -2 \binom{\frac{1}{2}(p-1)}{\frac{1}{8}(p-1)} \bmod p.$$

In each case, the claimed congruence is a mere rearrangement away. $\qquad\square$

In return, Theorem 3.22 has the following consequence in terms of Jacobsthal sums.

Corollary 3.24. *Let a be an integer. The following hold.*

(i) *If $p \equiv 1 \bmod 4$, then*

$$\sum_{x=0}^{p-1} \left(\frac{x^3 + ax}{p} \right) \equiv 2A_2(p)a^{(p-1)/4} \bmod p.$$

(ii) *If $p \equiv 1 \bmod 6$, then*

$$\sum_{x=0}^{p-1} \left(\frac{x^3 + a}{p} \right) \equiv 2A_3(p)a^{(p-1)/3} \bmod p.$$

Proof. In both cases, we adapt the strategy of Lemma 3.23. We use Euler's identity, and Corollary 1.2.

(i) Assume $p \equiv 1 \bmod 4$. We have

$$\sum_{x=0}^{p-1} \left(\frac{x^3 + ax}{p} \right) \equiv \sum_{x=0}^{p-1} (x^3 + ax)^{(p-1)/2} \bmod p$$

and

$$(x^3 + ax)^{(p-1)/2} = \sum_{k=0}^{(p-1)/2} \binom{\frac{1}{2}(p-1)}{k} a^{(p-1)/2-k} \, x^{(p-1)/2+2k}.$$

The only index for which $(p-1)/2+2k$ is a multiple of $p-1$ is $k = (p-1)/4$. Therefore,

$$\sum_{x=0}^{p-1} \left(\frac{x^3 + ax}{p} \right) \equiv -\binom{\frac{1}{2}(p-1)}{\frac{1}{4}(p-1)} a^{(p-1)/4} \equiv 2A_2(p)a^{(p-1)/4} \bmod p,$$

where, in the last step, we have used part (i) of Theorem 3.22.

(ii) Assume $p \equiv 1 \bmod 6$. We now have

$$\sum_{x=0}^{p-1} \left(\frac{x^3 + a}{p} \right) \equiv \sum_{x=0}^{p-1} (x^3 + a)^{(p-1)/2} \bmod p$$

and

$$(x^3 + a)^{(p-1)/2} = \sum_{k=0}^{(p-1)/2} \binom{\frac{1}{2}(p-1)}{k} a^{(p-1)/2-k} \, x^{(p-1)/2+3k}.$$

There is just one index for which $(p-1)/2 + 3k$ is a positive multiple of $p - 1$, and that is $k = (p-1)/6$. Hence,

$$\sum_{x=0}^{p-1} \left(\frac{x^3 + a}{p} \right) \equiv -\binom{\frac{1}{2}(p-1)}{\frac{1}{6}(p-1)} a^{(p-1)/3} \equiv 2A_3(p)a^{(p-1)/3} \bmod p$$

by using part (ii) of Theorem 3.22. $\qquad\square$

One use of the above corollary is that it can settle, in principle, some of the sign ambiguities in Theorem 3.17. The following example illustrates this idea.

Example 3.25. Let $p = 29$. We evaluate the sum of Legendre symbols

$$\sum_{x=0}^{p-1} \left(\frac{x^3 + 2x}{p} \right),$$

that is to say $\varphi_2(2)$. We note that $p \equiv 1 \bmod 4$; its representation as a sum of squares is $p = A^2 + B^2$, where $A = A_2(p) = -5$ and $B = \pm 2$. As $p \equiv 5 \bmod 8$, we know that 2 is a quadratic non-residue mod p. Thus, by part (i) of Theorem 3.17, we have $\varphi_2(2) = \pm 2B = \pm 4$.

On the other hand, part (i) of Corollary 3.24 gives

$$\varphi_2(2) \equiv 2A_2(p) \cdot 2^{(p-1)/4} = -10 \cdot 2^7 \equiv -4 \bmod p.$$

We deduce that $\varphi_2(2) = -4$.

Exercises

Exercise 3.26. Let $p > 3$ and fix $b \in \mathbb{F}_p$. For how many choices of $a \in \mathbb{F}_p$ is the polynomial $X^3 + aX + b$ irreducible over $\mathbb{F}_p$?

Exercise 3.27. Let $p \equiv 1 \bmod 2n$. Evaluate $\varphi_n(a)$ in the case $n = (p-1)/2$.

Exercise 3.28. Let $p \equiv 1 \bmod 6$. Consider the integers C and D, defined as the sums of Legendre symbols

$$C = \sum_{x=0}^{p-1} \left(\frac{x^3 + 1}{p} \right), \qquad D = \sum_{x=0}^{p-1} \left(\frac{a}{p} \right) \left(\frac{x^3 + a}{p} \right),$$

where a is a cubic non-residue modulo p, that is to say, a is an integer such that $x^3 \equiv a \bmod p$ has no solutions.

Show that $3p = C^2 + CD + D^2$.

Exercise 3.29. Let $a \in \mathbb{F}_p^*$. Prove that

$$|1 + \varphi_3(a)| \le 2\sqrt{p}, \quad |1 + \psi_4(a)| \le 2\sqrt{p}.$$

Exercise 3.30. Consider a cubic map of the form $f(x) = x^3 + a$, where $a \in \mathbb{F}_p^*$. Discuss the possibility that

$$\sum_{x \in \mathbb{F}_p} \sigma(f(x)) \equiv -1 \bmod p.$$

Exercise 3.31. Let $p \equiv 1 \bmod 2n$.

(i) Let $c \in \mathbb{F}_p^*$, and assume that $c \notin (\mathbb{F}_p^*)^n$. Show that

$$\sum_{a \in \mathbb{F}_p} \varphi_n(a)\varphi_n(ca) = 0.$$

(ii) Deduce that the w-sequence, defined by (70), satisfies

$$\sum_{k=0}^{n-1} w_k w_{k+s} = \begin{cases} -n(p-1) & \text{if } n \text{ is odd} \\ 0 & \text{if } n \text{ is even} \end{cases}$$

    for all $s = 1, \ldots, n-1$.

Exercise 3.32. For which primes p is it the case that, in $\mathbb{F}_p$, a run of three consecutive squares is as likely as a run of three consecutive non-squares?

Exercise 3.33. Let $p > 3$. Evaluate the double sum of Legendre symbols

$$\sum_{x,y=0}^{p-1} \left(\frac{x^6 - y^6}{p} \right).$$

Exercise 3.34. (i) Let j and k be integers satisfying $0 \le j \le k \le p-1$. Show that

$$\binom{p-1-j}{p-1-k} \equiv (-1)^{k-j} \binom{k}{j} \bmod p.$$

(ii) Let $p \equiv 1 \bmod 8$. Evaluate

$$\binom{\frac{7}{8}(p-1)}{\frac{1}{2}(p-1)} \bmod p.$$

Exercise 3.35. (i) Show that, for $k = 0, \ldots, \frac{1}{2}(p-1)$,

$$\binom{2k}{k} \equiv (-4)^k \binom{\frac{1}{2}(p-1)}{k} \bmod p.$$

(ii) Let $p \equiv 1 \bmod 8$. Evaluate

$$\binom{\frac{1}{4}(p-1)}{\frac{1}{8}(p-1)} \bmod p.$$

Notes

- Jacobsthal sums were introduced by Jacobsthal in his thesis (*Anwendungen einer Formel aus der Theorie der quadratischen Reste*, Dissertation, Berlin, 1906), written under Georg Frobenius and Issai Schur. The main point was to give a constructive realization of Fermat's sum-of-squares theorem, that is to say, an explicit representation $p = A^2 + B^2$ when $p \equiv 1$ mod 4. This is the content of Theorem 3.16 (i). This part of Jacobsthal's thesis was soon published as such (*Über die Darstellung der Primzahlen der Form $4n + 1$ als Summe zweier Quadrate*, J. Reine Angew. Math. 1907). Another one of Jacobsthal's goals was to count consecutive triples having a given pattern of squareness mod p. This is the context in which Lemma 3.19 appears in Jacobsthal's work; without this background, the proof looks magical!

  The name of Ernst Jacobsthal is probably unknown to most mathematicians. A glimpse of the man and his life comes from the memorial speech of Sigmund Selberg (*Ernst Jacobsthal*, Norske Vid. Selsk. Forh. (Trondheim) 1965; available in English translation at http://www.numb ertheory.org/obituaries/OTHERS/jacobsthal_eng.html). Jacobsthal was one among the many mathematicians having to flee Germany in 1939; see also the account of Siegmund-Schultze (*Mathematicians fleeing from Nazi Germany. Individual fates and global impact*, Princeton University Press 2009) in this regard.

- The explicit representation $p = A^2 + 3B^2$ when $p \equiv 1$ mod 6, described in part (ii) of Theorem 3.16, was given soon after Jacobsthal's work by von Schrutka (*Ein Beweis für die Zerlegbarkeit der Primzahlen von der Form $6n+1$ in ein einfaches und ein dreifaches Quadrat*, J. Reine Angew. Math. 1911). A variation was rediscovered by Chowla (*A formula similar to Jacobsthal's for the explicit value of x in $p = x^2 + y^2$ where p is a prime of the form $4k + 1$*, Proc. Lahore Philos. Soc. 1945). Both von Schrutka and Chowla employ φ_3-sums but, as we noted, one can choose to work with ψ_3-sums instead. The analogous formula using the ψ_3-sums was found independently by Emil Artin in 1921. According to the account of Roquette (*The Riemann Hypothesis in Characteristic p in Historical Perspective*, Springer 2018), the formula appears in a letter from Artin to his advisor, Gustav Herglotz.

  Exercise 3.28 is a minor variation on the Jacobsthal-type formula for primes of the form $p \equiv 1$ mod 6 case. This result was published fairly recently by Chan, Long, and Yang (*A cubic analogue of the Jacobsthal identity*, Amer. Math. Monthly 2011). Their argument actually runs, for the most part, on Jacobi sums. An argument based on Jacobsthal sums is presented in a recent note by Williams (*An arithmetic proof of a theorem of Chan, Long, and Yang*, Amer. Math. Monthly 2023).

- The explicit representation $p = A^2 + 2B^2$ when $p \equiv 1 \mod 8$, described in part (iii) of Theorem 3.16, is due to Whiteman (*Theorems analogous to Jacobsthal's theorem*, Duke Math. J. 1949; *Cyclotomy and Jacobsthal sums*, Amer. J. Math. 1952). In fact, the unified approach to Theorem 3.16–a triptych of theorems, really–largely follows that of Whiteman's papers. The notation for the w-sequence is chosen as a small tribute. It seems to us, in hindsight, that Jacobsthal's dissertation comes very close to distilling Theorem 3.16.

  An explicit representation $p = A^2 + 2B^2$ when $p \equiv 3 \mod 8$ seems unachievable by means of Jacobsthal sums only. Such a representation can be established, however, in terms of other quadratic character sums, see Hashimoto, Long, and Yang (*Jacobsthal identity for* $\mathbb{Q}(\sqrt{-2})$, Forum Math. 2012). The viewpoint in this latter paper is somewhat sophisticated. It would be desirable to have an elementary argument.

- Rankin (*Generalized Jacobsthal sums and sums of squares*, Acta Arith. 1987) has used generalized Jacobsthal sums, involving multiplicative characters of higher order, to obtain further explicit representations of p as sums of squares.

- The bounds of Corollary 3.12 are related to some celebrated bounds for quadratic character sums, due to Hasse (*Beweis des Analogons der Riemannschen Vermutung für die Artinschen und F.K. Schmidtschen Kongruenzzetafunktionen in gewissen zyklischen Fällen*, Nach. Gessel. Wiss. Göttingen Math.-Phys. Klasse 1933) and Weil (*Sur les courbes algébriques et les variétés qui s'en déduisent*, Hermann 1948). Consider a general quadratic character sum

$$\Sigma(f) = \sum_{x \in \mathbb{F}_p} \sigma(f(x)),$$

where f is a non-constant polynomial map with coefficients in $\mathbb{F}_p$. Broadly speaking, the Hasse–Weil bounds could be thought of as saying that

$$|\Sigma(f)| \leq (\deg(f) - 1)\sqrt{p}.$$

The bounds are, however, slighty more specific. Let us assume that f is square-free, in the sense that there is no non-constant polynomial h such that f is divisible by h^2. Hasse's results, addressing the "elliptic" case when f has low degree, settled a conjecture originally made by Emil Artin. The *Hasse bound* say that

$$\begin{cases} |\Sigma(f)| \leq 2\sqrt{p} & \text{if } \deg(f) = 3, \\ |1 + \Sigma(f)| \leq 2\sqrt{p} & \text{if } \deg(f) = 4. \end{cases} \tag{88}$$

Weil pioneered a more sophisticated approach which, in particular, yields a generalization of (88) in arbitrary degree. The *Weil bound* implies that

$$\begin{cases} |\Sigma(f)| \leq (\deg(f) - 1)\sqrt{p} & \text{if } \deg(f) \text{ is odd}, \\ |1 + \Sigma(f)| \leq (\deg(f) - 2)\sqrt{p} & \text{if } \deg(f) \text{ is even}. \end{cases} \tag{89}$$

For the lowest degree in each parity, $\deg(f) = 1$ respectively $\deg(f) = 2$, the bound (89) holds by (5), respectively, Theorem 2.4. These cases are essentially trivial. For the next lowest degree in each parity, $\deg(f) = 3$ respectively $\deg(f) = 4$, the bound (89) is the Hasse bound (88).

We thus see that the bounds of Corollary 3.12 are consistent with the Hasse–Weil bounds. In fact, the general bound (69) can be stronger, on par, or slightly weaker than the bound (89). The latter situation is illustrated by the following comparison: (69) only yields that $|\varphi_3(a)| \leq 3\sqrt{p}$ and $|\psi_4(a)| \leq 4\sqrt{p}$ for all $a \in \mathbb{F}_p^*$, whereas (88) predicts the sharper estimates $|1 + \varphi_3(a)| \leq 2\sqrt{p}$ and $|1 + \psi_4(a)| \leq 2\sqrt{p}$ for all $a \in \mathbb{F}_p^*$. But these improvements can also be obtained directly from our results on Jacobsthal sums; this is what Exercise 3.29 proposes.

- The sign ambiguities in Theorem 3.17 were resolved by Hudson and Williams (*Resolution of ambiguities in the evaluation of cubic and quartic Jacobsthal sums*, Pacific J. Math. 1982; *An application of a formula of Western to the evaluation of certain Jacobsthal sums*, Acta Arith. 1982). Soon after, Evans (*Determinations of Jacobsthal sums*, Pacific J. Math. 1984) gave a uniform treatment, based on Jacobi sums, to the problem of evaluating the Jacobsthal sums φ_n for $n = 2, 3, 4, 6, 10, 12$. See also Chapter 6 of the monograph *Gauss and Jacobi sums*, by Berndt, Evans, and Williams.

- The beautiful congruences for binomial coefficients established in Theorem 3.22 start with Gauss, who proved (i); (iii) is due to Jacobi and Stern. These results predate Jacobsthal sums, so the original approach was rather different. On the other hand, (ii) appears to be due to von Schrutka, who approached it via Jacobsthal sums in essentially the same way as we did it herein.

Chapter 4

Solution Counting with Jacobsthal Sums

In this chapter, we showcase a number of equations whose solution counting involves Jacobsthal sums of low degree. Notably, $\varphi_2(1)$ and $\varphi_3(1)$ appear quite often; we recall that

$$\varphi_2(1) = \begin{cases} 2A_2(p) & \text{if } p \equiv 1 \bmod 4, \\ 0 & \text{if } p \not\equiv 1 \bmod 4, \end{cases}$$

$$\varphi_3(1) = \begin{cases} 2A_3(p) - 1 & \text{if } p \equiv 1 \bmod 6, \\ -1 & \text{if } p \not\equiv 1 \bmod 6. \end{cases}$$

An assortment of examples

Example 4.1 (The last entry). Consider the equation

$$(1 + x^2)(1 + y^2) = 2. \tag{90}$$

For each $x \in \mathbb{F}_p$ such that $x^2 \neq -1$, we have

$$y^2 = -\frac{x^2 - 1}{x^2 + 1}$$

which has $1 + \sigma(-(x^2 - 1)(x^2 + 1)) = 1 + \sigma(-1)\sigma(x^4 - 1)$ solutions. Whence, the number of solutions to (90) is

$$N = \sum_{x^2 \neq -1} \left(1 + \sigma(-1)\sigma(x^4 - 1)\right)$$

$$= p - \left(1 + \sigma(-1)\right) + \sigma(-1)\psi_4(-1).$$

Recall that $\psi_4(-1) = -1 + \varphi_2(-1) = -1 + (-1)^{(p-1)/4}\varphi_2(1)$. Thus,

$$N = \begin{cases} p - 3 + (-1)^{(p-1)/4} \cdot 2A_2(p) & \text{if } p \equiv 1 \bmod 4, \\ p + 1 & \text{if } p \equiv 3 \bmod 4. \end{cases}$$

Example 4.2. We count the number of solutions to the equation

$$(x^2 + y^2)^2 = (1 - x^4)(1 - y^4). \tag{91}$$

This expands to $x^4 + y^4 + 2x^2y^2 = 1 + x^4y^4 - x^4 - y^4$, which can then be written as

$$2(x^2 + y^2)^2 = (1 + x^2y^2)^2. \tag{92}$$

The case when $x^2 + y^2 = 1 + x^2y^2 = 0$ amounts to $x^2 = 1$ and $y^2 = -1$, or $x^2 = -1$ and $y^2 = 1$; each case has $2(1 + \sigma(-1))$ solutions, whence a total of $4(1 + \sigma(-1))$ solutions.

If 2 is not a square, then there are no other solutions to (92). This degenerate situation occurs when $p \equiv 3, 5 \bmod 8$.

Assume $p \equiv 1, 7 \bmod 8$ in what follows. Then 2 is a square, say $j^2 = 2$. Now (92) amounts to two equations: $j(x^2+y^2) = 1+x^2y^2$, and $-j(x^2+y^2) = 1 + x^2y^2$. Let $N(j)$ and $N(-j)$ denote their respective number of solutions. The number of solutions to the Equation (92) is then

$$N = N(j) + N(-j) - 4(1 + \sigma(-1)), \tag{93}$$

where the correction term offsets the double counting of the solutions to $x^2 + y^2 = 1 + x^2y^2 = 0$.

We turn to evaluating $N(j)$. We write $j(x^2 + y^2) = 1 + x^2y^2$ as

$$y^2(x^2 - j) = jx^2 - 1.$$

Whenever $x^2 \neq j$, this equation has $1+\sigma((x^2-j)(jx^2-1))$ solutions. When $x^2 = j$ there are no solutions; for $x^2 = j$ and $jx^2 = 1$ would imply $j^2 = 1$, a contradiction. Thus,

$$N(j) = \sum_{x^2 \neq j} \left(1 + \sigma((x^2 - j)(jx^2 - 1))\right)$$

$$= p - (1 + \sigma(j)) + \sigma(j) \sum_{x \in \mathbb{F}_p} \sigma\left((x^2 - j)(x^2 - j^{-1})\right).$$

In general, for distinct $a, b \in \mathbb{F}_p$, the downgrading formula gives

$$\sum_{x \in \mathbb{F}_p} \sigma\left((x^2 - a)(x^2 - b)\right) = \sum_{x \in \mathbb{F}_p} \sigma\left(x(x - a)(x - b)\right) + \sum_{x \in \mathbb{F}_p} \sigma\left((x - a)(x - b)\right),$$

and we note that the latter sum equals -1, by (36). So,

$$\sum_{x \in \mathbb{F}_p} \sigma\left((x^2 - j)(x^2 - j^{-1})\right) = -1 + \sum_{x \in \mathbb{F}_p} \sigma\left(x(x - j)(x - j^{-1})\right).$$

Fortunately, the latter sum can be evaluated for the specific j at hand– the point being that $j^{-1} = j/2$. We are led to the change of variable $x := j(x + 1)/2$, which gives

$$\sum_{x \in \mathbb{F}_p} \sigma\left(x(x - j)\left(x - \frac{j}{2}\right)\right) = \sum_{x \in \mathbb{F}_p} \sigma\left(\frac{j(x + 1)}{2} \cdot \frac{j(x - 1)}{2} \cdot \frac{jx}{2}\right)$$

$$= \sigma(j)\varphi_2(-1),$$

since, we recall, 2 is a square. In summary

$$N(j) = p - (1 + \sigma(j)) + \sigma(j)(-1 + \sigma(j)\varphi_2(-1))$$

$$= p - 1 - 2\sigma(j) + \varphi_2(-1).$$

The analogous formula holds for $N(-j)$, as well. By (93), the number of solutions to the Equation (92) is

$$N = 2(p - 1) - 2\sigma(j)(1 + \sigma(-1)) + 2\varphi_2(-1) - 4(1 + \sigma(-1))$$

$$= \begin{cases} 2(p - 1) & \text{if } p \equiv 7 \bmod 8, \\ 2(p - 1) - 4\sigma(j) + 2\varphi_2(-1) - 8 & \text{if } p \equiv 1 \bmod 8. \end{cases}$$

In the case $p \equiv 1 \bmod 8$, we can be even more explicit. Firstly, we have $\varphi_2(-1) = (-1)^{(p-1)/4}\varphi_2(1) = \varphi_2(1) = 2A_2(p)$. Secondly,

$$\sigma(j) = (-1)^{(A_4(p)+1)/4}$$

as we have already seen in Example 3.21. In conclusion,

$$
N = \begin{cases} 2(p-1) & \text{if } p \equiv 7 \bmod 8, \\ 2(p-1) + 4A_2(p) - 4(-1)^{(A_4(p)+1)/4} - 8 & \text{if } p \equiv 1 \bmod 8. \end{cases}
$$

Example 4.3. Let $c \in \mathbb{F}_p^*$, and consider the equation

$$
x^2 + y^2 + z^2 + t^2 = 2cxyzt. \tag{94}
$$

We fix x and y, and we count the number of solutions (z,t). Set $\alpha(x,y) = cxy$ and $\beta(x,y) = x^2 + y^2$. Then Equation (94) reads $z^2 + t^2 + \beta = 2\alpha zt$, which we rewrite as

$$
(1 - \alpha^2)z^2 + (t - \alpha z)^2 = -\beta.
$$

If $\alpha^2 = 1$ then there are $(1 + \sigma(-\beta))p$ solutions. If $\alpha^2 \neq 1$ then, by Theorem 1.16, there are $p - \sigma(\alpha^2 - 1)$ solutions if $\beta \neq 0$, respectively, $p + \sigma(\alpha^2 - 1)(p - 1)$ solutions if $\beta = 0$. A formula that captures the number of solutions in all cases is

$$
p - \sigma(\alpha^2 - 1) + [\![\alpha^2 = 1]\!]\sigma(-\beta)p + [\![\beta = 0]\!]\sigma(\alpha^2 - 1)p.
$$

The number of solutions to the Equation (94) is then

$$
N = \sum_{x,y \in \mathbb{F}_p} \left(p - \sigma(\alpha^2 - 1) + [\![\alpha^2 = 1]\!]\sigma(-\beta)p + [\![\beta = 0]\!]\sigma(\alpha^2 - 1)p \right)
$$

$$
= p^3 - S_1 + pS_2 + pS_3,
$$

where

$$
S_1 = \sum_{x,y \in \mathbb{F}_p} \sigma(\alpha^2 - 1), \quad S_2 = \sum_{\substack{x,y \in \mathbb{F}_p \\ \alpha^2 = 1}} \sigma(-\beta), \quad S_3 = \sum_{\substack{x,y \in \mathbb{F}_p \\ \beta = 0}} \sigma(\alpha^2 - 1).
$$

We now compute each one of the three sums. Firstly,

$$
S_1 = \sum_{x \in \mathbb{F}_p} \sum_{y \in \mathbb{F}_p} \sigma((cx)^2 y^2 - 1) = \sigma(-1)p - (p-1),
$$

for the inner sum evaluates as $\sigma(-1)p$ when $x = 0$, and as $-\sigma((cx)^2) = -1$ when $x \neq 0$.

Secondly,

$$S_2 = \sigma(-1) \sum_{\substack{x,y \in \mathbb{F}_p \\ cxy = \pm 1}} \sigma(x^2 + y^2) = 2\sigma(-1) \sum_{x \in \mathbb{F}_p^*} \sigma(x^2 + c^{-2}x^{-2})$$

$$= 2\sigma(-1) \sum_{x \in \mathbb{F}_p^*} \sigma(x^4 + c^{-2}) = 2\sigma(-1)\big(\psi_4(c^{-2}) - 1\big).$$

Thirdly,

$$S_3 = \sum_{\substack{x,y \in \mathbb{F}_p \\ y^2 = -x^2}} \sigma(c^2 x^2 y^2 - 1).$$

The term $x = 0$ contributes $\sigma(-1)$. For each $x \neq 0$, there are $1 + \sigma(-1)$ choices for y. Thus,

$$S_3 = \sigma(-1) + (1 + \sigma(-1)) \sum_{x \in \mathbb{F}_p^*} \sigma(-c^2 x^4 - 1)$$

$$= \sigma(-1) + (1 + \sigma(-1))\sigma(-1) \sum_{x \in \mathbb{F}_p^*} \sigma(x^4 + c^{-2})$$

$$= \sigma(-1) + (1 + \sigma(-1))\big(\psi_4(c^{-2}) - 1\big).$$

In summary,

$$N = p^3 + p - 1 + \big(1 + 3\sigma(-1)\big)\big(\psi_4(c^{-2}) - 1\big)p.$$

We recall that $\psi_4(c^{-2}) = -1 + \varphi_2(c^{-2}) = -1 + \sigma(c)\varphi_2(1)$. So, to conclude, the number of solutions to Equation (94) is

$$N = p^3 + p - 1 + \big(1 + 3\sigma(-1)\big)\big(\sigma(c)\varphi_2(1) - 2\big)p$$

$$= \begin{cases} p^3 - 1 + \big(8\sigma(c)A_2(p) - 7\big)p & \text{if } p \equiv 1 \bmod 4, \\ p^3 - 1 + 5p & \text{if } p \equiv 3 \bmod 4. \end{cases}$$

Example 4.4. Let $a \in \mathbb{F}_p^*$. Consider the elliptic equation

$$y^2 = x^3 + ax, \tag{95}$$

whose number of solutions we denote by N_a. As we know, the solution count is closely related to a Jacobsthal sum, namely,

$$N_a = p + \varphi_2(a).$$

When $p \equiv 3 \bmod 4$, the sum φ_2 vanishes identically on $\mathbb{F}_p^*$. Thus, $N_a = p$ for each $a \in \mathbb{F}_p^*$.

Consider now the case when $p \equiv 1 \bmod 4$. By Theorem 3.17, the possible values of $\varphi_2(a)$ are $\pm 2A$ and $\pm 2B$, where $p = A^2 + B^2$. This relation clearly precludes both integers A and B from vanishing, so $\varphi_2(a) \neq 0$ for each $a \in \mathbb{F}_p^*$. Whence $N_a \neq p$ for each $a \in \mathbb{F}_p^*$.

Let us then consider the following question: how close can N_a be to p, and under what circumstances do such "near-p" solution counts occur? As the sum φ_2 takes even values, the potential near-p values for N_a are $p \pm 2$. We discuss in detail how these are achieved. Once again, $\varphi_2(a)$ equals $\pm 2A$ or $\pm 2B$, where $p = A^2 + B^2$ and A is odd while B is even. Then $\varphi_2(a) = \pm 2$ implies $A = \pm 1$; so necessarily p is of the form $4m^2 + 1$, for some integer m.

Conversely, if p is of the form $4m^2 + 1$, then $A = \pm 1$ and $B = \pm 2m$. We infer that $\varphi_2(a) = \pm 2$ when $a \in \mathbb{F}_p^*$ is a square, respectively, $\varphi_2(a) = \pm 4m$ otherwise. When a is a square, say $a = c^2$, we can be even more precise. Since $A_2(p) = -1$, we know that $\varphi_2(1) = -2$; so $\varphi_2(a) = \sigma(c)\varphi_2(1) = -2\sigma(c)$. Thus both $N_a = p - 2$ and $N_a = p + 2$ occur equally often, and they are achieved as follows: the former precisely when $a \in \mathbb{F}_p^*$ is a fourth power, the latter precisely when $a \in \mathbb{F}_p^*$ is a square but not a fourth power.

Solution counting for $x^3 + y^3$ and $x^4 + y^4$

In this section, we consider the Fermat-type equation

$$x^n + y^n = c,$$

where $c \in \mathbb{F}_p$. We give solution counts in the cubic case $n = 3$, and in the quartic case $n = 4$. The linear case, $n = 1$, is obvious while the quadratic case, $n = 2$, is covered by Theorem 1.16. Recall from Example 1.15 that we may assume $p \equiv 1 \bmod n$.

Theorem 4.5. *Assume $p \equiv 1 \bmod 3$. Then the number of solutions to the equation*

$$x^3 + y^3 = c \tag{96}$$

is given as follows:

$$\begin{cases} p - 1 + \varphi_3(c/2) & \text{if } c \neq 0, \\ 3p - 2 & \text{if } c = 0. \end{cases}$$

Proof. The case $c = 0$ is easily settled. The equation $x^3 + y^3 = 0$ admits the solution $(0,0)$, while the non-zero solutions are in bijection with the solutions to $x^3 = -1$, $y \in \mathbb{F}_p^*$, by the change of variable $x := yx$. Thus, the number of solutions to $x^3 + y^3 = 0$ is

$$1 + (p-1) \cdot \#\{x : x^3 = -1\} = 1 + 3(p-1) = 3p - 2.$$

Consider now the main case $c \neq 0$. We cancel the cubic y term by the change of variables $x := x - y$; the Equation (96) becomes

$$x^3 - 3x^2 y + 3xy^2 = c.$$

Necessarily, $x \neq 0$. The latter equation is quadratic in y, with discriminant $\Delta = (-3x^2)^2 - 4 \cdot 3x(x^3 - c) = -3x^4 + 12cx$, so it has $1 + \sigma(-3x^4 + 12cx)$ solutions for each $x \neq 0$. Whence the total number of solutions to (96) is

$$\sum_{x \in \mathbb{F}_p^*} \left(1 + \sigma(-3x^4 + 12cx)\right) = p - 1 + \sigma(-3) \sum_{x \in \mathbb{F}_p^*} \sigma(x^4 - 4cx)$$

$$= p - 1 + \varphi_3(-4c)$$

since $\sigma(-3) = 1$. Using periodicity and sign change properties for the φ_3-sum, namely (62) and (63), we have $\varphi_3(-4c) = \varphi_3(4c) = \varphi_3(c/2)$. We conclude that, for $c \neq 0$, the number of solutions to (96) is $p - 1 + \varphi_3(c/2)$. $\qquad\square$

Example 4.6. When $p \equiv 1 \bmod 3$, the equation

$$x^3 + y^3 = 2 \tag{97}$$

has $p - 2 + 2A_3(p)$ solutions.

Example 4.7. Assume $p \equiv 1 \bmod 3$, and consider the equation

$$x^3 + y^3 + cz^3 = 0, \tag{98}$$

where $c \in \mathbb{F}_p^*$. The number of solutions, denoted by $N_3(c)$, is given by

$$N_3(c) = N(x^3 + y^3 = 0) + (p-1)N(x^3 + y^3 = -c).$$

Using Theorem 4.5, we find that

$$N_3(c) = (3p - 2) + (p-1)\big(p - 1 + \varphi_3(-c/2)\big)$$

$$= p^2 + p - 1 + (p-1)\varphi_3(c/2).$$

In the last step, we used the sign symmetry of the φ_3-sum.

In particular, the equation $x^3 + y^3 + 2z^3 = 0$ has $p^2 + (p-1) \cdot 2A_3(p)$ solutions.

We now turn to the quartic Fermat equation.

Theorem 4.8. *Assume $p \equiv 1 \bmod 4$. Then the number of solutions to the equation*

$$x^4 + y^4 = c \tag{99}$$

is given as follows:

$$\begin{cases} p - 1 - 2\epsilon + 2\epsilon\varphi_2(c) + \sigma(c)\varphi_2(1) & \text{if } c \neq 0, \\ 1 + (2 + 2\epsilon)(p - 1) & \text{if } c = 0, \end{cases}$$

where $\epsilon = (-1)^{(p-1)/4}$.

Proof. We first consider the solution count for $c = 0$. The equation $x^4 + y^4 = 0$ has $1 + (p-1) \cdot \#\{x : x^4 = -1\}$ solutions; indeed, the case $y = 0$ yields one solution, while the case $y \neq 0$ allows for a change of variables $x := yx$. The equation $x^4 = -1$ has 0 or 4 solutions, according to whether or not -1 is a fourth power; in other words, $2 + 2\epsilon$ solutions. In summary, the number of solutions to $x^4 + y^4 = 0$ is $1 + (2 + 2\epsilon)(p - 1)$.

Consider now the case $c \neq 0$. Let $j \in \mathbb{F}_p$ with $j^2 = -1$. Then (99) can be rewritten as

$$(x^2 + jy^2)(x^2 - jy^2) = c.$$

We are led to consider, for each $s \in \mathbb{F}_p^*$, the system $x^2 + jy^2 = s$, $x^2 - jy^2 = s^{-1}c$. These equations amount to the system

$$x^2 = \frac{s + s^{-1}c}{2} = \frac{s^2 + c}{2s}, \qquad y^2 = \frac{s - s^{-1}c}{2j} = \frac{s^2 - c}{2js}$$

whose number of solutions is

$$\left(1 + \sigma\big(2s(s^2 + c)\big)\right)\left(1 + \sigma\big(2js(s^2 - c)\big)\right).$$

We have $\sigma(j) = j^{(p-1)/2} = \epsilon$ and $\sigma(2) = (-1)^{(p^2-1)/8} = \epsilon$ as well. Thus $\sigma(2j) = 1$. We deduce that the number of solutions to (99) is

$$\sum_{s \in \mathbb{F}_p^*} \left(1 + \epsilon\sigma(s(s^2 + c))\right)\left(1 + \sigma(s(s^2 - c))\right)$$

$$= p - 1 + \epsilon \sum_{s \in \mathbb{F}_p^*} \sigma(s(s^2 + c)) + \sum_{s \in \mathbb{F}_p^*} \sigma(s(s^2 - c)) + \epsilon \sum_{s \in \mathbb{F}_p^*} \sigma(s^4 - c^2).$$

The three sums are all Jacobsthal sums. The first one is $\varphi_2(c)$. The second sum is $\varphi_2(-c)$, and we note that $\varphi_2(-c) = \epsilon\varphi_2(c)$. The third sum is $\psi_4(-c^2)-\sigma(-c^2) = \psi_4(-c^2)-1$, and we note that $\psi_4(-c^2) = \varphi_2(-c^2)-1 = \epsilon\sigma(c)\varphi_2(1) - 1$. Overall, the displayed expression evaluates as

$$p - 1 + 2\epsilon\varphi_2(c) + \epsilon(\epsilon\sigma(c)\varphi_2(1) - 2) = p - 1 - 2\epsilon + 2\epsilon\varphi_2(c) + \sigma(c)\varphi_2(1),$$

as claimed. $\square$

Example 4.9. The number of solutions to the equation

$$x^4 + y^4 = 1 \tag{100}$$

is

$$\begin{cases} p - 3 + 6A_2(p) & \text{if } p \equiv 1 \bmod 8, \\ p + 1 - 2A_2(p) & \text{if } p \equiv 5 \bmod 8. \end{cases}$$

Example 4.10. Assume $p \equiv 1 \bmod 4$, and consider the equation

$$x^4 + y^4 + cz^4 = 0, \tag{101}$$

where $c \in \mathbb{F}_p^*$. The number of solutions, denoted by $N_4(c)$, is given by

$$N_4(c) = N(x^4 + y^4 = 0) + (p - 1)N(x^4 + y^4 = -c).$$

By Theorem 4.8, we have

$$N(x^4 + y^4 = 0) = 1 + (2 + 2\epsilon)(p - 1)$$

and

$$N(x^4 + y^4 = -c) = p - 1 - 2\epsilon + 2\epsilon\varphi_2(-c) + \sigma(-c)\varphi_2(1)$$
$$= p + 1 - (2 + 2\epsilon) + 2\varphi_2(c) + \sigma(c)\varphi_2(1)$$

as $\sigma(-c) = \sigma(c)$, and $\varphi_2(-c) = \epsilon\varphi_2(c)$. A pleasant calculation leads us to

$$N_4(c) = p^2 + \big(2\varphi_2(c) + \sigma(c)\varphi_2(1)\big)(p - 1).$$

In particular if c is a square, say $c = a^2$, then $\varphi_2(c) = \sigma(a)\varphi_2(1)$. We then obtain the following explicit solution count:

$$N_4(a^2) = p^2 + \big(1 + 2\sigma(a)\big) \cdot 2A_2(p) \cdot (p - 1).$$

Solution counting for $x^4 + y^4 + z^4 + t^4$

Theorem 4.11. *Let $p \equiv 1 \bmod 4$. Then the number of solutions to the equation*

$$x^4 + y^4 + z^4 + t^4 = c \tag{102}$$

is given as follows, where $\epsilon = (-1)^{(p-1)/4}$:

- *if $c = 0$,*

$$p^3 + (5 + 12\epsilon)(p-1)p + \varphi_2(1)^2(p-1);$$

- *if $c \in \mathbb{F}_p^*$ is not a square,*

$$p^3 - (5 + 12\epsilon)p + (8 + 4\epsilon)\varphi_2(c)p - (6 + 4\epsilon)\varphi_2(1)p - \varphi_2(1)^2;$$

- *if $c = d^2$ for some $d \in \mathbb{F}_p^*$,*

$$p^3 - (5 + 12\epsilon)p + \big((16 + 4\epsilon)\sigma(d) + (6 + 4\epsilon)\big)\varphi_2(1)p - \varphi_2(1)^2.$$

Proof. We break the proof into three steps.

Step 1. By the separation principle, the solution-counting map for $x^4 + y^4 + z^4 + t^4$ is given by the convolution formula

$$N_{x^4+y^4+z^4+t^4} = N_{x^4+y^4} * N_{z^4+t^4} = (N_{x^4+y^4})^{*2}.$$

The latter notation stands for the square of $N_{x^4+y^4}$ under convolution. The solution-counting map $N_{x^4+y^4}$ is evaluated in Theorem 4.8; the following functional formula can be easily deduced:

$$N_{x^4+y^4} = p\mathbb{1} + (1 + 2\epsilon)\kappa + 2\epsilon\varphi_2 + \varphi_2(1)\sigma.$$

We start computing

$$(N_{x^4+y^4})^{*2} = \big(p\mathbb{1} + (1 + 2\epsilon)\kappa + 2\epsilon\varphi_2 + \varphi_2(1)\sigma\big)^{*2}$$

$$= p^3\mathbb{1} + \big((1 + 2\epsilon)\kappa + 2\epsilon\varphi_2 + \varphi_2(1)\sigma\big)^{*2}.$$

Indeed, $\mathbb{1}^{*2} = p\mathbb{1}$, and $\mathbb{1} * \alpha = 0$ whenever α is a map on $\mathbb{F}_p$ with vanishing total sum; as κ, σ, and φ_2 have this property, so does any of their integral combinations. In the same vein, $\kappa * \alpha = (p\delta_0 - \mathbb{1}) * \alpha = p\alpha$ whenever α is an arithmetic map on $\mathbb{F}_p$ with vanishing total sum. Therefore, $N_{x^4+y^4+z^4+t^4} = (N_{x^4+y^4})^{*2}$ equals

$$N_{x^4+y^4+z^4+t^4} = p^3\mathbb{1} + (1+2\epsilon)^2 p\kappa + 2(1+2\epsilon)p\big(2\epsilon\varphi_2 + \varphi_2(1)\sigma\big)$$

$$+ \big(2\epsilon\varphi_2 + \varphi_2(1)\sigma\big)^{*2}. \tag{103}$$

Next, we would like to expand the latter square. By Lemma 1.13, and the fact that $\sigma(-1) = 1$, we know that $\sigma * \sigma = \kappa$. We need to evaluate $\sigma * \varphi_2$ and $\varphi_2 * \varphi_2$.

Step 2. We claim that

$$(\varphi_2 * \sigma)(c) = \begin{cases} 2\epsilon\sigma(d)p & \text{if } c = d^2 \text{ for some } d \in \mathbb{F}_p^* \\ 0 & \text{otherwise.} \end{cases} \tag{104}$$

For each $c \in \mathbb{F}_p$, we have

$$(\varphi_2 * \sigma)(c) = \sum_{a+b=c} \varphi_2(a)\sigma(b) = \sum_{a+b=c} \sum_{x\in\mathbb{F}_p} \sigma\big(x(x^2+a)\big)\sigma(b)$$

$$= \sum_{x\in\mathbb{F}_p} \sigma(x) \sum_{a+b=c} \sigma(x^2+a)\sigma(b).$$

The inner sum is $(\sigma * \sigma)(x^2 + c) = \kappa(x^2 + c) = p[\![x^2 + c = 0]\!] - 1$. Whence

$$(\varphi_2 * \sigma)(c) = \sum_{x\in\mathbb{F}_p} \sigma(x)\big(p[\![x^2 = -c]\!] - 1\big) = p \sum_{x^2=-c} \sigma(x).$$

Let $\Sigma_1(c)$ denote the right-most sum. Clearly, $\Sigma_1(0) = 0$. If c is not a square in $\mathbb{F}_p^*$, then nor is $-c$ and so $\Sigma_1(c) = 0$. If c is a square in $\mathbb{F}_p^*$, say $c = d^2$, then $x^2 = -c$ admits two solutions: $x = \pm jd$, where $j^2 = -1$. In this case $\Sigma_1(c) = \sigma(jd) + \sigma(-jd) = 2\epsilon\sigma(d)$ since $\sigma(j) = \epsilon$, as pointed out in proof of Theorem 4.5. The formula for $\varphi_2 * \sigma$ is thereby checked.

Next, we claim that

$$(\varphi_2 * \varphi_2)(c) = \begin{cases} \big(\sigma(c)\varphi_2(1) - 2\epsilon\big)p & \text{if } c \neq 0 \\ 2\epsilon(p-1)p & \text{if } c = 0. \end{cases} \tag{105}$$

For each $c \in \mathbb{F}_p$, we have:

$$(\varphi_2 * \varphi_2)(c) = \sum_{a+b=c} \varphi_2(a)\varphi_2(b) = \sum_{a+b=c} \sum_{x,y\in\mathbb{F}_p} \sigma\big(x(x^2+a)\big)\sigma\big(y(y^2+b)\big)$$

$$= \sum_{x,y\in\mathbb{F}_p} \sigma(xy) \sum_{a+b=c} \sigma(x^2+a)\sigma(y^2+b).$$

The inner sum is $(\sigma*\sigma)(x^2+y^2+c) = \kappa(x^2+y^2+c) = p[\![x^2+y^2+c=0]\!]-1$. Whence

$$(\varphi_2 * \varphi_2)(c) = \sum_{x,y\in\mathbb{F}_p} \sigma(xy)\big(p[\![x^2+y^2=-c]\!]-1\big) = p \sum_{x^2+y^2=-c} \sigma(xy).$$

Thus, in order to justify (105), it suffices to check that

$$\sum_{x^2+y^2=-c} \sigma(xy) = \begin{cases} \sigma(c)\varphi_2(1) - 2\epsilon & \text{if } c \neq 0, \\ 2\epsilon(p-1) & \text{if } c = 0. \end{cases}$$

Consider the case when $c = 0$. For each $x \in \mathbb{F}_p^*$, the equation $x^2+y^2 = 0$ has two solutions $y = \pm jx$. Correspondingly, $\sigma(xy) = \sigma(\pm jx^2) = \sigma(j) = \epsilon$. Whence

$$\sum_{x^2+y^2=0} \sigma(xy) = 2\epsilon(p-1).$$

Now let $c \neq 0$. For $x \in \mathbb{F}_p^*$, we make the change of variable $y := tx$. Thus,

$$\sum_{x^2+y^2=-c} \sigma(xy) = \sum_{x^2(1+t^2)=-c} \sigma(t).$$

Consider the constraint $x^2(1+t^2) = -c$, viewed as an equation in x; it has no solutions when $t^2+1 = 0$, respectively, $1+\sigma(-c/(t^2+1)) = 1+\sigma(c)\sigma(t^2+1)$ solutions when $t^2 + 1 \neq 0$. We obtain

$$\sum_{x^2(1+t^2)=-c} \sigma(t) = \sum_{1+t^2\neq 0} \sigma(t)\big(1 + \sigma(c)\sigma(t^2+1)\big)$$

$$= \sum_{t\in\mathbb{F}_p} \sigma(t) + \sigma(c) \sum_{t\in\mathbb{F}_p} \sigma(t)\sigma(t^2+1) - \sum_{1+t^2=0} \sigma(t)$$

$$= \sigma(c)\varphi_2(1) - \big(\sigma(j) + \sigma(-j)\big) = \sigma(c)\varphi_2(1) - 2\epsilon.$$

This completes the proof of (105).

Step 3. Returning to (103), we expand it into the somewhat formidable formula

$$N_{x^4+y^4+z^4+t^4} = p^3\mathbb{1} + (5+4\epsilon)p\kappa + (8+4\epsilon)p\varphi_2 + (2+4\epsilon)p\varphi_2(1)\sigma$$
$$+ 4(\varphi_2 * \varphi_2) + 4\epsilon\varphi_2(1)(\varphi_2 * \sigma) + \varphi_2(1)^2\kappa.$$

We evaluate at $c = 0$, respectively, at $c \neq 0$, making use of (104) and (105). We get

$$N_{x^4+y^4+z^4+t^4}(0) = p^3 + (5+4\epsilon)p(p-1) + 8\epsilon(p-1)p + \varphi_2(1)^2(p-1)$$
$$= p^3 + (5+12\epsilon)p(p-1) + \varphi_2(1)^2(p-1).$$

As for $c \neq 0$, the outcome is

$$N_{x^4+y^4+z^4+t^4}(c) = p^3 - (5+4\epsilon)p + (8+4\epsilon)\varphi_2(c)p + (2+4\epsilon)\sigma(c)\varphi_2(1)p$$
$$+ 4\big(\sigma(c)\varphi_2(1) - 2\epsilon\big)p + 4\epsilon\varphi_2(1)(\varphi_2 * \sigma)(c) - \varphi_2(1)^2$$
$$= p^3 - (5+12\epsilon)p + (8+4\epsilon)\varphi_2(c)p + (6+4\epsilon)\varphi_2(1)\sigma(c)p$$
$$+ 4\epsilon\varphi_2(1)(\varphi_2 * \sigma)(c) - \varphi_2(1)^2.$$

Further splitting into two more cases, c a square, respectively, not a square in $\mathbb{F}_p^*$, leads to the claimed formulas. Phew. $\qquad\square$

We leave it as an exercise for the reader to check that the formulas we have obtained in Theorem 4.11 for the solution-counting map $N_{x^4+y^4+z^4+t^4}$ do satisfy the identity

$$\sum_{c\in\mathbb{F}_p} N_{x^4+y^4+z^4+t^4}(c) = p^4.$$

Example 4.12. For $p \equiv 1 \bmod 4$, the number of solutions to

$$x^4 + y^4 + z^4 + t^4 = 1$$

is $p^3 - (5+12\epsilon)p + (44+16\epsilon)A_2(p)p - 4A_2(p)^2$.

Lehmers' equation

Let $p > 3$. We count the number of solutions in $\mathbb{F}_p^*$ for the equation

$$\frac{1}{x} + \frac{1}{y} + \frac{1}{z} = x + y + z. \qquad (106)$$

The equation looks pleasing in its symmetry, unassuming in its simplicity. Alas, the solution count will turn out to be rather intricate, and it will require a few tricks.

We proceed in three steps.

Step 1. The change of variables $x := xz$, $y := yz$ turns (106) into the equation

$$z^2(1 + x + y) = 1 + \frac{1}{x} + \frac{1}{y}.$$

Splitting into two cases, according to whether $1 + x + y = 0$ or not, we see that the latter equation has $N = n_0 + n_1$ solutions, where

$$n_0 = (p-1) \cdot \# \left\{ (x,y) \in \mathbb{F}_p^* \times \mathbb{F}_p^* : 1 + x + y = 1 + \frac{1}{x} + \frac{1}{y} = 0 \right\}$$

$$= (p-1) \cdot \# \left\{ x \in \mathbb{F}_p^* : 1 + \frac{1}{x} - \frac{1}{x+1} = 0 \right\} = (p-1)(1 + \sigma(-3)),$$

and

$$n_1 = \sum_{\substack{x,y \in \mathbb{F}_p^* \\ x+y+1 \neq 0}} \left(1 + \sigma(1 + x + y)\sigma\left(1 + \frac{1}{x} + \frac{1}{y}\right) \right)$$

$$= (p-1)^2 - (p-2) + \sum_{x,y \in \mathbb{F}_p^*} \sigma(1 + x + y)\sigma\left(1 + \frac{1}{x} + \frac{1}{y}\right).$$

Overall, we find that

$$N = (p-1)^2 + 1 + (p-1)\sigma(-3) + \Sigma, \tag{107}$$

where

$$\Sigma = \sum_{x,y \in \mathbb{F}_p^*} \sigma(1 + x + y)\sigma\left(1 + \frac{1}{x} + \frac{1}{y}\right). \tag{108}$$

Step 2. The spotlight is now on the sum Σ. We invert one of the variables, say $y := 1/y$, and we rewrite Σ as follows:

$$\Sigma = \sum_{x,y \in \mathbb{F}_p^*} \sigma\left(1 + x + \frac{1}{y}\right)\sigma\left(1 + y + \frac{1}{x}\right)$$

$$= \sum_{x,y \in \mathbb{F}_p^*} \sigma\left(y + \frac{y+1}{x}\right)\sigma\left(x + \frac{x+1}{y}\right).$$

Now, we make the rational change of variables

$$u = \frac{x+1}{y}, \qquad v = \frac{y+1}{x}$$

with inverse

$$x = \frac{u+1}{uv-1}, \qquad y = \frac{v+1}{uv-1}. \tag{109}$$

These birational transformations establish a bijective correspondence between the uv-domain $\{(u,v) : u, v \neq -1, uv \neq 1\}$ and the xy-domain $\{(x,y) : x, y \neq 0, x+y+1 \neq 0\}$. As

$$y + \frac{y+1}{x} = \frac{uv^2+1}{uv-1}, \qquad x + \frac{x+1}{y} = \frac{u^2v+1}{uv-1},$$

we obtain

$$\sum_{\substack{x,y \neq 0 \\ x+y+1 \neq 0}} \sigma\left(y + \frac{y+1}{x}\right) \sigma\left(x + \frac{x+1}{y}\right) = \sum_{\substack{u,v \neq -1 \\ uv \neq 1}} \sigma(uv^2+1)\sigma(u^2v+1).$$

$$\tag{110}$$

The left-hand side of (110) accounts for much of the sum Σ. The missing piece is

$$\sum_{\substack{x,y \neq 0 \\ x+y+1=0}} \sigma\left(y + \frac{y+1}{x}\right) \sigma\left(x + \frac{x+1}{y}\right) = \sum_{\substack{x,y \neq 0 \\ x+y+1=0}} \sigma(y-1)\sigma(x-1)$$

$$= \sigma(-1) \sum_{x \neq 0,-1} \sigma(x+2)\sigma(x-1)$$

$$= \sigma(-1)(-1 - 2\sigma(-2)).$$

On the right-hand side of (110), we can get rid of the condition $uv \neq 1$ in a similar way, by computing

$$\sum_{\substack{u,v \neq -1 \\ uv=1}} \sigma(uv^2+1)\sigma(u^2v+1) = \sum_{u \neq 0,-1} \sigma\left(\frac{1}{u}+1\right)\sigma(u+1)$$

$$= \sum_{u \neq 0,-1} \sigma(u) = -\sigma(-1).$$

We deduce from (110) that

$$\Sigma = -2\sigma(2) + \sum_{u,v\neq-1} \sigma(uv^2+1)\sigma(u^2v+1).$$

Actually, we can pursue the same idea by noting that

$$\sum_{\substack{u\neq-1 \\ v=-1}} \sigma(uv^2+1)\sigma(u^2v+1) = \sum_{u\neq-1} \sigma(u+1)\sigma(1-u^2)$$

$$= \sum_{u\neq-1} \sigma(1-u) = -\sigma(2)$$

and similarly for the sum over $u = -1$, $v \neq -1$. These two sums have one common term, corresponding to $u = v = -1$, and that term vanishes. We conclude, neatly, that

$$\Sigma = \sum_{u,v\in\mathbb{F}_p} \sigma(uv^2+1)\sigma(u^2v+1). \tag{111}$$

Step 3. Set

$$\Sigma^* = \sum_{u,v\in\mathbb{F}_p^*} \sigma(uv^2+1)\sigma(u^2v+1), \tag{112}$$

so that

$$\Sigma = 2p - 1 + \Sigma^*. \tag{113}$$

If $p \equiv 2 \bmod 3$ then the system $uv^2 = X$, $u^2v = Y$ admits a unique solution for any $X,Y \in \mathbb{F}_p^*$; that solution is $u = YZ^{-1}$, $v = XZ^{-1}$, where $Z \in \mathbb{F}_p^*$ is uniquely defined by $Z^3 = XY$. In this case,

$$\Sigma^* = \sum_{X,Y\in\mathbb{F}_p^*} \sigma(X+1)\sigma(Y+1)$$

$$= \left(\sum_{X\in\mathbb{F}_p^*} \sigma(X+1)\right)\left(\sum_{Y\in\mathbb{F}_p^*} \sigma(Y+1)\right) = 1.$$

Therefore, $\Sigma = 2p$, and

$$N = p^2 + 2 + (p-1)\sigma(-3) = p^2 - p + 3. \tag{114}$$

Consider now the case when $p \equiv 1 \mod 3$. As $(uv^2)(u^2v) = (uv)^3$ is in $(\mathbb{F}_p^*)^3$, a subgroup of index 3 in the multiplicative group $\mathbb{F}_p^*$, we can write

$$uv^2 = g^k X^3, \quad u^2 v = g^{-k} Y^3,$$

where $X, Y \in \mathbb{F}_p^*$, g is a generator of $\mathbb{F}_p^*$, and $k \in \{0, 1, 2\}$. In fact, given X, Y and k, there are exactly three solutions to the above system–namely $u = g^{-k} Y^3 (\tau XY)^{-1}$, $v = g^k X^3 (\tau XY)^{-1}$ where $\tau \in \mathbb{F}_p^*$ runs over the three solutions to the equation $\tau^3 = 1$. In this case, we have

$$\Sigma^* = \frac{1}{3} \sum_{k=0}^{2} \sum_{X,Y \in \mathbb{F}_p^*} \sigma(g^k X^3 + 1)\sigma(g^{-k} Y^3 + 1)$$

$$= \frac{1}{3} \sum_{k=0}^{2} \left(\sum_{X \in \mathbb{F}_p^*} \sigma(X^3 + g^{-k}) \right) \left(\sum_{Y \in \mathbb{F}_p^*} \sigma(Y^3 + g^k) \right).$$

The first parenthesized sum is $\psi_3(g^{-k}) - \sigma(g^k)$; by the inversion identity (60), this equals $\sigma(g^k)\varphi_3(g^k)$. Similarly, the second parenthesized sum equals $\sigma(g^{-k})\varphi_3(g^{-k})$. Thus,

$$\Sigma^* = \frac{1}{3} \sum_{k=0}^{2} \varphi_3(g^k)\varphi_3(g^{-k}).$$

It is convenient to write the above formula in terms of the w-sequence, given by $w_k = \varphi_3(g^k)$. Keeping in mind the 3-periodicity, specifically that $w_{-k} = w_{3-k}$, we have

$$\Sigma^* = \frac{1}{3}\left(w_0^2 + 2w_1 w_2\right).$$

We have seen in the proof of Theorem 3.16 that

$$w_0^2 + w_1^2 + w_2^2 = 6p + 3, \qquad w_0 + w_1 + w_2 = -3.$$

Squaring the relation $w_1 + w_2 = -(w_0 + 3)$, and using the first relation, we arrive at the expression $w_0^2 + 2w_1 w_2 = 3w_0^2 + 6w_0 + 6 - 6p$; therefore

$$\Sigma^* = (1 + w_0)^2 + 1 - 2p.$$

It follows that $\Sigma = (1 + w_0)^2 = (1 + \varphi_3(1))^2 = 4A_3(p)^2$. As $\sigma(-3) = 1$, we obtain

$$N = p^2 - p + 1 + 4A_3(p)^2. \tag{115}$$

In summary, the number of solutions to the Equation (106) is given as follows:

$$
\begin{cases}
p^2 - p + 1 + 4A_3(p)^2 & \text{if } p \equiv 1 \bmod 3, \\
p^2 - p + 3 & \text{if } p \equiv 2 \bmod 3.
\end{cases}
$$

Exercises

Exercise 4.13. Let $c \in \mathbb{F}_p^*$. Count the number of solutions to the equation

$$
x(2y^2 - c) = y(x^2 - c).
$$

Exercise 4.14. Let $p \equiv 1 \bmod 4$. Count the number of solutions to the equation

$$
x^4 + y^4 + z^4 = -1.
$$

Exercise 4.15. Let $p \equiv 1 \bmod 3$, and $c \in \mathbb{F}_p^*$. Count the number of solutions to the equation

$$
x^3 + y^3 + c(z^3 + t^3) = 0,
$$

and to the equation

$$
x^3 + c(y^3 + z^3) = 1.
$$

Exercise 4.16. Let $p > 29$. Show that the equation $x^4 + y^4 = c$ is solvable, for every $c \in \mathbb{F}_p$. Discuss the solvability when $p = 29$.

Exercise 4.17. Let $a \in \mathbb{F}_p^*$ and $c \in \mathbb{F}_p$. Count the number of solutions to the equation

$$
x^4 + y^4 + az^2 = c.
$$

Exercise 4.18. Show that every element in $\mathbb{F}_p$ can be written as the product of three elements whose sum vanishes, with the exception of two specific elements in a specific $\mathbb{F}_p$.

Exercise 4.19. Let $p > 3$, and $c \in \mathbb{F}_p^*$. Count the number of solutions to the equation

$$
\frac{1}{x} + \frac{1}{y} + \frac{1}{z} = c(x + y + z).
$$

Exercise 4.20. (i) Show that

$$\sum_{x,y\in\mathbb{F}_p} \sigma\big(xy(xy+1)(x+y)\big) = \begin{cases} 4A_2(p)^2 - 2p & \text{if } p \equiv 1 \bmod 4, \\ 0 & \text{if } p \not\equiv 1 \bmod 4. \end{cases}$$

(ii) Count the number of solutions to the equation

$$z^2 = (x^2 + y^2)(1 + x^2 y^2).$$

Exercise 4.21. (i) Show that

$$\sum_{x,y\in\mathbb{F}_p} \sigma\big(xy(x+1)(y+1)(x-y)\big) = \sum_{u,v\in\mathbb{F}_p} \sigma\big(uv(uv+1)(u+v)\big).$$

(ii) Count the number of solutions to the equation

$$w^2 = (x^2 - y^2)(y^2 - z^2)(z^2 - x^2).$$

Notes

- The equation of Example 4.1 can also be written as $x^2 + y^2 + x^2 y^2 = 1$. The solution count for this equation modulo p appears in the last entry, dated 1814, of Gauss's mathematical notebook. The notebook is a private record that Gauss kept of his mathematical discoveries; the first entry, from 1796, concerns the constructibility of the regular 17-gon. Gauss arrived at the above equation by studying the arc-length of the lemniscate–the plane curve given by the equation $(x^2 + y^2)^2 = x^2 - y^2$. There are lemniscatic analogues of the sine and cosine functions, denoted sl and cl, and they turn out to satisfy the functional relation $\mathrm{sl}^2 + \mathrm{cl}^2 + \mathrm{sl}^2 \mathrm{cl}^2 = 1$. This is what really brought Gauss to the equation $x^2 + y^2 + x^2 y^2 = 1$. The solution count for the equation $x^2 + y^2 + x^2 y^2 = 1$ over $\mathbb{F}_p$ by means of quadratic character sums is due to Chowla (*The last entry in Gauss's diary*, Proc. Nat. Acad. Sci. U.S.A. 1940).

- Example 4.3 is the simplest case of a more general result due to Carlitz (*Certain special equations in a finite field*, Monatsh. Math. 1954).

- Primes of the form $am^2 + bm + c$ crop up in several places–Example 4.4, the solutions to Exercise 3.30 and Exercise 3.32. A prominent open problem, first raised by Euler though often referred to as *Landau's conjecture*, asks whether there are infinitely many primes of the form $4m^2 + 1$. More generally, the *Hardy–Littlewood "Conjecture F"* proposes that there are infinitely many primes of the form $am^2 + bm + c$ whenever the coefficients $a, b, c \in \mathbb{Z}$ are subject to some obvious requirements: a, b, c are relatively prime; $a + b$ and c are not both even; $b^2 - 4ac$ is not a square.

As we have seen in Example 4.4, for primes of the form $p = 4m^2 + 1$ the Jacobsthal sums $\varphi_2(a)$ can achieve near-vanishing values. At the same time, the Jacobsthal sums $\varphi_2(a)$ nearly attain the values $\pm 2\sqrt{p}$; specifically, $\varphi_2(a) = \pm 4m = \pm 2\sqrt{p-1}$ whenever $a \in \mathbb{F}_p^*$ is not a square. This shows that the Hasse bound for cubics (88) is essentially sharp.

- Solution counts for the cubic equations $x^3 + y^3 = c$ and $x^3 + y^3 + cz^3 = 0$ were first obtained by Gauss, though his formulas take slightly different forms than those given in Theorem 4.5 and Example 4.7. Further results were obtained by Chowla, Cowles, and Cowles (*On the number of zeros of diagonal cubic forms*, J. Number Theory 1977; *The number of zeroes of $x^3 + y^3 + cz^3$ in certain finite fields*, J. Reine Angew. Math. 1978; *On the difference of cubes* (mod p), Acta Arith. 1980). Neither one of these references uses Jacobsthal sums.

  Comprehensive, though more technical, solution counts for the cubic and quartic diagonal equations, $a_1 x_1^3 + \cdots + a_n x_n^3 = c$ and $a_1 x_1^4 + \cdots + a_n x_n^4 = c$, can be found in Chapter 10 of the monograph *Gauss and Jacobi sums*, by Berndt, Evans, and Williams. Their approach relies on generalized Jacobi sums.

- The Equation (106) is our framing of a result due to Lehmer and Lehmer (*On the cubes of Kloosterman sums*, Acta Arith. 1960). Taking a cue from the old masters–Euler, Gauss, Jacobi–the Lehmers distilled many of their number theoretical discoveries from extensive computations. For the particular facts under consideration, they make an interesting recording of their odyssey:

  > These results were discovered empirically by an inspection of numerical results in 1952. After repeatedly unsuccessful attempts over the intervening years, a proof of these formulas was completed in 1959.

  The solution count for Equation (106) is a lengthy, multi-step argument. The most intricate step is the evaluation of the sum Σ. Mordell (*On Lehmer's congruence associated with cubes of Kloosterman's sums*, J. London Math. Soc. 1961), whose streamlined proof we follow to a large extent, writes that he "would have thought it very difficult indeed to evaluate $[\Sigma]$ in simple terms". The main piece of magic, discovered and used by the Lehmers in a slightly different form, is the

xy-to-uv change of variables in Step 2. Mordell's simplification is mainly in Step 3: whereas the Lehmers' original argument introduces certain double Jacobsthal sums, Mordell skilfully relates to the usual Jacobsthal sums. The Lehmers actually needed the general form of Equation (106), exhibited in Exercise 4.19. But that is not hard to obtain, once (106) has been cracked.

Lehmers' xy-to-uv change of variables is also used, quite transparently, in part (i) of Exercise 4.21. This exercise is a variation on an argument due, independently, to Thomason (*On finite Ramsey numbers*, European J. Combin. 1982) and to Evans, Pulham, and Sheehan (*On the number of complete subgraphs contained in certain graphs*, J. Combin. Theory Ser. B 1981). The problem that they considered is that of counting the number of copies of the complete graph on four vertices in the Paley graph of order p, where $p \equiv 1 \bmod 4$.

- Exercise 4.18 is a result of Klyachko and Vassilyev (*Balanced factorizations*, Amer. Math. Monthly 2016). They resort to the Hasse bound, which seems too powerful a tool for the problem at hand.

- Part (ii) of Exercise 4.20 recovers a solution count obtained, by different means, in a very recent work of Kiritchenko, Tsfasman, Vlăduţ, and Zakharevich (*Quadratic residue patterns, algebraic curves and a K3 surface*, Finite Fields Appl. 2025).

Chapter 5

Further Quadratic Character Sums

Introducing the ϱ-sums

We have already discussed two families of quadratic character sums whose argument is a cubic polynomial–the Jacobsthal sums

$$\varphi_2(c) = \sum_{x \in \mathbb{F}_p} \sigma(x^3 + cx), \qquad \psi_3(c) = \sum_{x \in \mathbb{F}_p} \sigma(x^3 + c).$$

We now highlight another family of quadratic character sums over cubics. We define the ϱ-*sum* with parameter $c \in \mathbb{F}_p$ by

$$\varrho(c) = \sum_{x \in \mathbb{F}_p} \sigma(x^3 + x^2 + cx). \tag{116}$$

The ϱ-sums account for most quadratic character sums whose arguments are reducible cubics. Indeed, up to discarding the leading coefficient and translating the variable if necessary, a reducible cubic polynomial takes the form $f(x) = x^3 + bx^2 + cx$. Then

$$\sum_{x \in \mathbb{F}_p} \sigma(x^3 + bx^2 + cx) = \sigma(b)\varrho\left(\frac{c}{b^2}\right) \quad \text{when } b \neq 0, \tag{117}$$

after a change of variable $x := bx$ in the left-hand sum. Note that the case $b = 0$ is the Jacobsthal sum $\varphi_2(c)$. However, the usefulness of ϱ-sums goes well beyond reducible cubics; as we will see in subsequent sections, quadratic character sums over certain quartics and quintics can be expressed in terms of ϱ-sums. In turn, a handful of ϱ-sums can be made explicit by means of Jacobsthal sums.

We start by recording some immediate evaluations of ϱ-sums.

Theorem 5.1. *We have*

$$\varrho(0) = -1, \qquad \varrho\left(\frac{1}{4}\right) = -\sigma(-2). \tag{118}$$

For $p > 3$, we have

$$\varrho\left(\frac{1}{3}\right) = \psi_3(1), \qquad \varrho\left(\frac{2}{9}\right) = \sigma(6)\varphi_2(1). \tag{119}$$

Proof. The evaluations in (118) are direct:

$$\varrho(0) = \sum_{x \in \mathbb{F}_p} \sigma(x^2(x+1)) = \sum_{x \in \mathbb{F}_p^*} \sigma(x+1) = -\sigma(1) = -1;$$

$$\varrho\left(\frac{1}{4}\right) = \sum_{x \in \mathbb{F}_p} \sigma\left(x\left(x+\frac{1}{2}\right)^2\right) = \sum_{x \neq -1/2} \sigma(x) = -\sigma\left(-\frac{1}{2}\right) = -\sigma(-2).$$

As for the evaluations in (119), they are somewhat indirect. We write

$$\psi_3(-1) = \sum_{x \in \mathbb{F}_p} \sigma(x^3 - 1) = \sum_{x \in \mathbb{F}_p} \sigma(x^3 + 3x^2 + 3x) = \sigma(3)\varrho\left(\frac{1}{3}\right);$$

$$\varphi_2(-1) = \sum_{x \in \mathbb{F}_p} \sigma(x^3 - x) = \sum_{x \in \mathbb{F}_p} \sigma(x^3 + 3x^2 + 2x) = \sigma(3)\varrho\left(\frac{2}{9}\right).$$

Both computations employ the change of variable $x := x+1$, and then (117). Furthermore, we have $\psi_3(-1) = \sigma(-1)\psi_3(1)$, and $\varphi_2(-1) = \sigma(2)\varphi_2(1)$. Thus,

$$\varrho\left(\frac{1}{3}\right) = \sigma(-3)\psi_3(1), \qquad \varrho\left(\frac{2}{9}\right) = \sigma(6)\varphi_2(1).$$

The first formula can be simplified further, observing that $\sigma(-3)\psi_3(1) = \psi_3(1)$. For $\psi_3(1) = 0$ when $p \equiv 2 \bmod 3$, whereas $\sigma(-3) = 1$ for $p \equiv 1 \bmod 3$. $\qquad\square$

We will often adopt the viewpoint that formulas for ϱ-sums in terms of the Jacobsthal sums φ_2 and ψ_3, such as the formulas in (119), are satisfactory evaluations. If needed, they can often be made much more explicit, as in the following example.

Example 5.2. Let $p > 3$, and consider the quadratic character sum

$$\sum_{x \in \mathbb{F}_p} \sigma(x^4 - 9x^2 + 27). \tag{120}$$

By the downgrading formula (31), the above character sum–which we denote by Σ in what follows–can be rewritten as

$$\Sigma = \sum_{x \in \mathbb{F}_p} \sigma(x^2 - 9x + 27) + \sum_{x \in \mathbb{F}_p} \sigma\big(x(x^2 - 9x + 27)\big).$$

The first sum evaluates to -1, since the quadratic argument has non-vanishing discriminant. The second sum evaluates as $\sigma(-1)\varrho(1/3)$, by (117). These two facts, combined with the evaluation of $\varrho(1/3)$, yield the explicit formula

$$\Sigma = -1 + \sigma(-1)\psi_3(1) = \begin{cases} \sigma(-1) \cdot 2A_3(p) - 1 & \text{if } p \equiv 1 \bmod 3, \\ -1 & \text{if } p \equiv 2 \bmod 3. \end{cases}$$

A twisted symmetry

Consider the following table, which collects the values of ϱ-sums for $p = 13$.

c	0	1	2	3	4	5	6	7	8	9	10	11	12
$\varrho(c)$	-1	2	4	2	6	0	-6	-2	-4	-2	1	2	-2

We might notice the single vanishing; the only two odd values; and what appears to be a signed pairing: 6 and -6, 4 and -4, and so on. It is quite easy to figure out a general explanation for the two odd values. Namely, odd values of ϱ-sums correspond to those parameters c for which the cubic $x^3 + x^2 + cx$ has repeated roots, that is, for $c = 0$ and $c = 1/4$; the actual values at these parameters are spelled out by (118).

The reason behind the signed pairing is an unexpected twisted symmetry of the ϱ-sums.

Theorem 5.3. *We have*

$$\varrho\left(\frac{1}{4} - c\right) = \sigma(-2)\varrho(c). \tag{121}$$

Proof. For $c = 0$, we indeed have $\varrho(1/4) = -\sigma(-2) = \sigma(-2)\varrho(0)$. Assume $c \neq 0$ in what follows. We write

$$\varrho(c) = \sum_{x \in \mathbb{F}_p} \sigma(x^3 + x^2 + cx) = \sum_{x \in \mathbb{F}_p^*} \sigma\left(x(x^2 + x + c)\right)$$

$$= \sum_{x \in \mathbb{F}_p^*} \sigma\left(\frac{x^2 + x + c}{x}\right) = \sum_{y \in \mathbb{F}_p} \sigma(y) \cdot \# \left\{ x \in \mathbb{F}_p^* : y = \frac{x^2 + x + c}{x} \right\}.$$

For $y \in \mathbb{F}_p$, consider $yx = x^2 + x + c$ as a quadratic equation in x. It has discriminant $(y-1)^2 - 4c = y^2 - 2y + 1 - 4c$, and the assumption $c \neq 0$ rules out $x = 0$ as a solution. Therefore,

$$\# \left\{ x \in \mathbb{F}_p^* : y = \frac{x^2 + x + c}{x} \right\} = 1 + \sigma(y^2 - 2y + 1 - 4c).$$

Consequently,

$$\varrho(c) = \sum_{y \in \mathbb{F}_p} \sigma(y)\left(1 + \sigma(y^2 - 2y + 1 - 4c)\right)$$

$$= \sum_{y \in \mathbb{F}_p} \sigma\left(y^3 - 2y^2 + (1 - 4c)y\right) = \sigma(-2)\varrho\left(\frac{1 - 4c}{4}\right)$$

thanks to (117). $\square$

By combining the twisted symmetry with the two evaluations in (119) we deduce two additional evaluations of ϱ-sums.

Corollary 5.4. *If $p \neq 3$, then*

$$\varrho\left(-\frac{1}{12}\right) = \sigma(-2)\psi_3(1), \qquad \varrho\left(\frac{1}{36}\right) = \sigma(-3)\varphi_2(1). \tag{122}$$

Example 5.5. Let $p > 3$, and consider the character sum

$$\sum_{x \in \mathbb{F}_p} \sigma(x^3 - 15x - 22). \tag{123}$$

The cubic argument is reducible since it admits $x = -2$ as a root. After the change of variable $x := x - 2$, we obtain

$$\sum_{x \in \mathbb{F}_p} \sigma(x^3 - 15x - 22) = \sum_{x \in \mathbb{F}_p} \sigma(x^3 - 6x^2 - 3x)$$

$$= \sigma(-6)\varrho\left(-\frac{1}{12}\right) = \sigma(3)\psi_3(1).$$

The twisted symmetry compels us to consider what happens for the parameter $c = 1/8$, the center of symmetry. The corresponding ϱ-sum is a so-called *Brewer sum*. The twisted symmetry reads $\varrho(1/8) = \sigma(-2)\varrho(1/8)$. As $\sigma(-2) = -1$ precisely when $p \equiv 5, 7 \bmod 8$, we deduce the following.

Corollary 5.6. *If $p \equiv 5, 7 \bmod 8$ then*

$$\varrho\left(\frac{1}{8}\right) = 0. \tag{124}$$

Quadratic character sums over split cubics

We now consider quadratic character sums whose argument is a special type of reducible cubics–namely, a split cubic.

Theorem 5.7. *Let $c \in \mathbb{F}_p$ with $c \neq -1$. Then*

$$\sum_{x \in \mathbb{F}_p} \sigma\big(x(x+1)(x+c)\big) = \sigma(-2(c+1))\varrho\left(\left(\frac{1}{2} \cdot \frac{c-1}{c+1}\right)^2\right). \tag{125}$$

Proof. Using (117), we have

$$\sum_{x \in \mathbb{F}_p} \sigma\big(x(x+1)(x+c)\big) = \sum_{x \in \mathbb{F}_p} \sigma\big(x^3 + (c+1)x^2 + cx\big)$$

$$= \sigma(c+1)\varrho\left(\frac{c}{(c+1)^2}\right).$$

Next, the twisted symmetry gives

$$\varrho\left(\frac{c}{(c+1)^2}\right) = \sigma(-2)\varrho\left(\frac{1}{4} - \frac{c}{(c+1)^2}\right) = \sigma(-2)\varrho\left(\left(\frac{1}{2} \cdot \frac{c-1}{c+1}\right)^2\right).$$

Formula (125) follows by combining the above facts. $\qquad\square$

Let us illustrate some uses of formula (125). We start with an intriguing twisted symmetry of ϱ-sums with square arguments.

Theorem 5.8. *Let $p > 3$. Consider the involution $\eta : \mathbb{F}_p\backslash\{1/6\} \to \mathbb{F}_p\backslash\{1/6\}$ given by*

$$\eta(c) = \frac{2c+1}{12c-2}.$$

Then for each $c \in \mathbb{F}_p\backslash\{1/6\}$ the following holds:

$$\varrho(\eta(c)^2) = \sigma(12c-2)\varrho(c^2). \tag{126}$$

Proof. The change of variable $x := -x - 1$ yields

$$\sum_{x\in\mathbb{F}_p} \sigma\big(x(x+1)(x+c)\big) = \sigma(-1) \sum_{x\in\mathbb{F}_p} \sigma\big(x(x+1)(x+1-c)\big)$$

as long as $c \neq -1, 2$. Applying formula (125) to both sides gives

$$\sigma(c+1)\varrho\left(\left(\frac{c-1}{2(c+1)}\right)^2\right) = \sigma(c-2)\varrho\left(\left(\frac{c}{2(c-2)}\right)^2\right).$$

This relation suggests that we consider the maps

$$\gamma(c) = \frac{c-1}{2(c+1)}, \qquad \delta(c) = \frac{c}{2(c-2)}.$$

Thus, $\gamma : \mathbb{F}_p\backslash\{-1\} \to \mathbb{F}_p\backslash\{1/2\}$ and $\delta : \mathbb{F}_p\backslash\{2\} \to \mathbb{F}_p\backslash\{1/2\}$. Both are bijections, with inverses given by

$$\gamma^{-1}(c) = \frac{1+2c}{1-2c}, \qquad \delta^{-1}(c) = \frac{4c}{2c-1}.$$

We also have $\gamma(2) = \delta(-1) = 1/6$. We may then consider the composition $\gamma \circ \delta^{-1}$, which maps $\mathbb{F}_p\backslash\{1/6, 1/2\}$ to itself, and it is given by

$$\gamma \circ \delta^{-1}(c) = \frac{2c+1}{12c-2}.$$

Thus, $\gamma \circ \delta^{-1}(c) = \eta(c)$ for $c \neq 1/6, 1/2$.

Now the change of variable $c := \delta^{-1}(c)$ turns the relation obtained above,

$$\sigma(c+1)\varrho(\gamma(c)^2) = \sigma(c-2)\varrho(\delta(c)^2) \qquad (c \neq -1, 2)$$

into

$$\sigma\big(\delta^{-1}(c) + 1\big)\varrho(\eta(c)^2) = \sigma\big(\delta^{-1}(c) - 2\big)\varrho(c^2) \qquad (c \neq 1/6, 1/2).$$

This can be rearranged as

$$\varrho(\eta(c)^2) = \sigma(12c - 2)\varrho(c^2) \qquad (c \neq 1/6, 1/2).$$

At $c = 1/2$, a fixed point of η, the above formula clearly holds as well. $\qquad\square$

Next, we consider another application in which formula (125) plays a key role.

Example 5.9. Consider the quadratic character sum

$$\sum_{x,y,z \in \mathbb{F}_p} \sigma\big(xyz(x + y)(y + z)(z + x)\big). \tag{127}$$

There is no contribution coming from $z = 0$. For $z \neq 0$, we make the change of variables $x := xz$, $y := yz$; the triple sum (127) becomes $(p - 1)\Sigma$ where

$$\Sigma := \sum_{x,y \in \mathbb{F}_p} \sigma\big(xy(x + 1)(y + 1)(x + y)\big). \tag{128}$$

Next, we write the double sum Σ as an iterated sum:

$$\Sigma = \sum_{x \in \mathbb{F}_p} \sigma\big(x(x + 1)\big) \sum_{y \in \mathbb{F}_p} \sigma\big(y(y + 1)(y + x)\big).$$

There is no contribution coming from $x = -1$. For $x \neq -1$, we may appeal to (125), leading us to the formula

$$\Sigma = \sigma(-2) \sum_{x \neq -1} \sigma(x)\varrho\left(\left(\frac{1}{2} \cdot \frac{x - 1}{x + 1}\right)^2\right).$$

We have already noted that the map $x \mapsto \frac{1}{2} \cdot \frac{x-1}{x+1}$ is a bijection from $\mathbb{F}_p \backslash \{-1\}$ to $\mathbb{F}_p \backslash \{\frac{1}{2}\}$, with inverse $t \mapsto \frac{1+2t}{1-2t}$. Therefore,

$$\sigma(-2)\Sigma = \sum_{t \neq 1/2} \sigma\left(\frac{1 + 2t}{1 - 2t}\right)\varrho(t^2) = \sum_{t \in \mathbb{F}_p} \sigma(1 - 4t^2)\varrho(t^2).$$

By invoking the downgrading formula (31), we can write

$$\sum_{t\in\mathbb{F}_p}\sigma(1-4t^2)\varrho(t^2) = \sum_{t\in\mathbb{F}_p}\sigma(1-4t)\varrho(t) + \sum_{t\in\mathbb{F}_p}\sigma\big(t(1-4t)\big)\varrho(t).$$

The first sum on the right-hand side is easily evaluated:

$$\sum_{t\in\mathbb{F}_p}\sigma(1-4t)\varrho(t) = \sum_{t\in\mathbb{F}_p}\sigma\left(\frac{1}{4}-t\right)\sum_{x\in\mathbb{F}_p}\sigma(x^3+x^2+tx)$$

$$= \sum_{x\in\mathbb{F}_p}\sigma(-x)\sum_{t\in\mathbb{F}_p}\sigma\left(t-\frac{1}{4}\right)\sigma(t+x+x^2)$$

$$= \sum_{x\in\mathbb{F}_p}\sigma(-x)\left(p\left[\!\!\left[x^2+x=-\frac{1}{4}\right]\!\!\right]-1\right)$$

$$= \sum_{x\in\mathbb{F}_p}\sigma(-x)\cdot p\left[\!\!\left[x=-\frac{1}{2}\right]\!\!\right] = \sigma(2)p.$$

As for the second sum on the right-hand side, we perform the change of variable $t := \frac{1}{4}-t$ to obtain

$$\sum_{t\in\mathbb{F}_p}\sigma\big(t(1-4t)\big)\varrho(t) = \sum_{t\in\mathbb{F}_p}\sigma\big(t(1-4t)\big)\varrho\left(\frac{1}{4}-t\right)$$

$$= \sigma(-2)\sum_{t\in\mathbb{F}_p}\sigma\big(t(1-4t)\big)\varrho(t),$$

where, in the last step, we have used once again the twisted symmetry of ϱ-sums. The upshot is that, whenever $\sigma(-2) = -1$, the sum under consideration vanishes.

We conclude that, for $p \equiv 5, 7 \bmod 8$, we have

$$\Sigma = \sigma(-1)p,$$

and so the triple sum (127) equals $\sigma(-1)p(p-1)$.

Quadratic character sums over some quartics I

The ϱ-sums are quadratic character sums with cubic arguments. In this section and the next, we discuss how the ϱ-sums also account for certain quadratic character sums with quartic arguments.

We start with a result that addresses certain reducible quartics–namely, those that are the product of two quadratics.

Theorem 5.10. *Let $g_1(x) = x^2 + b_1 x + c_1$ and $g_2(x) = x^2 + b_2 x + c_2$ be two distinct quadratics, at least one of which is not a square. Put*

$$\Delta_1 = b_1^2 - 4c_1, \qquad \Delta_2 = b_2^2 - 4c_2, \qquad B = 4(c_1 + c_2) - 2b_1 b_2.$$

Then

$$\sum_{x \in \mathbb{F}_p} \sigma\big(g_1(x)g_2(x)\big) = \begin{cases} -1 + \sigma(B)\varrho\left(\dfrac{\Delta_1\Delta_2}{B^2}\right) & \text{if } B \neq 0, \\[2mm] -1 + \varphi_2(\Delta_1\Delta_2) & \text{if } B = 0. \end{cases} \tag{129}$$

Proof. By hypothesis, at least one of the discriminants Δ_1 and Δ_2 is non-zero. If, say, $\Delta_1 = 0$ then $g_1(x) = (x + b_1/2)^2$ and $B = 4g_2(-b_1)$. We easily compute

$$\sum_{x \in \mathbb{F}_p} \sigma\big(g_1(x)g_2(x)\big) = \sum_{x \neq -b_1/2} \sigma\big(g_2(x)\big)$$

$$= -1 - \sigma\big(g_2(-b_1/2)\big) = -1 - \sigma(B),$$

which checks (129) in this case. On the way, we have used the evaluation $\sum_{x \in \mathbb{F}_p} \sigma\big(g_2(x)\big) = -1$; this holds thanks to the fact that $\Delta_2 \neq 0$.

We continue by assuming that $\Delta_1, \Delta_2 \neq 0$. We claim that

$$\sum_{x \in \mathbb{F}_p} \sigma\big(g_1(x)g_2(x)\big) = -1 + \sum_{y \in \mathbb{F}_p} \sigma(y)\sigma(\Delta_1 y^2 + By + \Delta_2). \tag{130}$$

Formula (129) will immediately follow by rewriting the latter sum: when $B = 0$, it equals $\sigma(\Delta_1)\varphi_2(\Delta_2/\Delta_1) = \varphi_2(\Delta_1\Delta_2)$; when $B \neq 0$, it can be brought to the form $\sigma(B)\varrho((\Delta_1\Delta_2)/B^2)$ thanks to (117).

We turn to checking (130). We first consider the case when $g_1(x)$ and $g_2(x)$ share a root in $\mathbb{F}_p$. In this case, the two quadratics take the form $g_1(x) = (x - x_0)(x - x_1)$, respectively, $g_2(x) = (x - x_0)(x - x_2)$, where x_0, x_1, x_2 are distinct. The left-hand side of (130) can be evaluated as follows:

$$\sum_{x \in \mathbb{F}_p} \sigma\big((x - x_0)^2(x - x_1)(x - x_2)\big) = \sum_{x \neq x_0} \sigma\big((x - x_1)(x - x_2)\big)$$

$$= -1 - \sigma\big((x_0 - x_1)(x_0 - x_2)\big).$$

On the other hand, it is easy to see that $\Delta_1 = (x_0 - x_1)^2$, $\Delta_2 = (x_0 - x_2)^2$, and $B = -2(x_0 - x_1)(x_0 - x_2)$. Hence,

$$\Delta_1 y^2 + By + \Delta_2 = \big(y(x_0 - x_1) - (x_0 - x_2)\big)^2$$

and so $\sigma(\Delta_1 y^2 + By + \Delta_2) = 1$, except when $y = (x_0 - x_2)/(x_0 - x_1)$. The right-hand side of (130) becomes

$$-1 + \sum_{y \neq (x_0 - x_2)/(x_0 - x_1)} \sigma(y) = -1 - \sigma\big((x_0 - x_2)/(x_0 - x_1)\big)$$

$$= -1 - \sigma\big((x_0 - x_1)(x_0 - x_2)\big).$$

The verification of (130) is complete in this case.

Next, we consider the case when $g_1(x)$ and $g_2(x)$ have no common roots in $\mathbb{F}_p$. Arguing as in the proof of Theorem 5.3, we write

$$\sum_{x \in \mathbb{F}_p} \sigma\big(g_1(x)g_2(x)\big) = \sum_{x \in \mathbb{F}_p \,:\, g_1(x) \neq 0} \sigma\left(\frac{g_2(x)}{g_1(x)}\right) = \sum_{y \in \mathbb{F}_p} \sigma(y) N(y),$$

where $N(y)$ is the number of solutions $x \in \mathbb{F}_p$ to the equation $y = g_2(x)/g_1(x)$. This equation is equivalent to $yg_1(x) = g_2(x)$, since $g_1(x)$ and $g_2(x)$ have no common roots; in turn, the latter equation takes the explicit form

$$(y - 1)x^2 + (b_1 y - b_2)x + (c_1 y - c_2) = 0. \tag{131}$$

If $y \neq 1$ then (131) is a true quadratic in x; the number of solutions is $1 + \sigma(\Delta(y))$, where

$$\Delta(y) = (b_1 y - b_2)^2 - 4(y - 1)(c_1 y - c_2) = \Delta_1 y^2 + By + \Delta_2.$$

For $y = 1$, (131) degenerates into a linear equation, $(b_1 - b_2)x + (c_1 - c_2) = 0$. If $b_1 \neq b_2$, there is one solution. If $b_1 = b_2$ then necessarily $c_1 \neq c_2$, since $g_1(x)$ and $g_2(x)$ are distinct; in this case there are no solutions. Thus $N(1) = [\![b_1 \neq b_2]\!]$. On the other hand, $\Delta(1) = (b_2 - b_1)^2$ so $\sigma(\Delta(1)) = [\![b_1 \neq b_2]\!]$. We summarize as follows:

$$N(y) = \begin{cases} 1 + \sigma(\Delta(y)) & \text{if } y \neq 1, \\ \sigma(\Delta(1)) & \text{if } y = 1. \end{cases}$$

Therefore,

$$\sum_{y\in\mathbb{F}_p} \sigma(y)N(y) = -1 + \sum_{y\in\mathbb{F}_p} \sigma(y)\big(1+\sigma(\Delta(y))\big)$$

$$= -1 + \sum_{y\in\mathbb{F}_p} \sigma(y)\sigma(\Delta_1 y^2 + By + \Delta_2),$$

and (130) is verified in this case, as well. $\square$

We illustrate Theorem 5.10 with the following concrete example. Further applications will arise in the next section.

Example 5.11. Consider the quadratic character sum

$$\sum_{x\in\mathbb{F}_p} \sigma(x^4 + 6x^3 - 6x - 1). \tag{132}$$

The quartic argument factorizes as $(x^2 - 1)(x^2 + 6x + 1)$. Here, $\Delta_1 = 4$, $\Delta_2 = 32$, and $B = 0$. Therefore,

$$\sum_{x\in\mathbb{F}_p} \sigma(x^4 + 6x^3 - 6x - 1) = -1 + \varphi_2(4\cdot 32) = -1 + \sigma(2)\varphi_2(2).$$

Applications to solution counting

Throughout this section, we assume that $p > 3$.

Example 5.12. We count the number of solutions to the equation

$$(1 + x^2)(1 + y^2) = -1. \tag{133}$$

In fact, let us consider, more generally, the equation

$$(1 + x^2)(1 + y^2) = c, \tag{134}$$

where $c \neq 0, 2$. We note that, for $c = 2$, this is the "last entry" equation discussed in Example 4.1. The above equation is equivalent to

$$y^2 = -\frac{x^2 + 1 - c}{x^2 + 1},$$

and so the number of solutions is given by

$$N(c) = \sum_{x \in \mathbb{F}_p \,:\, x^2 \neq -1} \left(1 + \sigma \left(-\frac{x^2 + 1 - c}{x^2 + 1} \right) \right)$$

$$= p - \left(1 + \sigma(-1) \right) + \sigma(-1) \sum_{x \in \mathbb{F}_p} \sigma\big((x^2 + 1)(x^2 + 1 - c) \big).$$

The latter sum can be evaluated by using Theorem 5.10. Alternatively, one can use the downgrading formula, and then Theorem 5.7. Either way we find that

$$N(c) = p - 1 - 2\sigma(-1) + \sigma(c - 2)\varrho \left(\frac{1 - c}{(2 - c)^2} \right).$$

The particular case at hand is $c = -1$; then $c \neq 2$, indeed, as $p \neq 3$. The above ϱ-sum is $\varrho(2/9) = \sigma(6)\varphi_2(1)$. We conclude that

$$N(-1) = p - 1 - 2\sigma(-1) + \sigma(-2)\varphi_2(1)$$

$$= \begin{cases} p - 3 + \sigma(2) \cdot 2A_2(p) & \text{if } p \equiv 1 \bmod 4, \\ p + 1 & \text{if } p \equiv 3 \bmod 4. \end{cases}$$

Example 5.13. For $c \in \mathbb{F}_p^*$, consider the equation

$$\left(2 + x + \frac{1}{x} \right) \left(2 + y + \frac{1}{y} \right) = c. \tag{135}$$

The number of solutions, which we denote by $N(c)$, can be thought of as being given by the formula

$$N(c) = \sum_{a,b \in \mathbb{F}_p \,:\, ab = c} \#\{x : 2 + x + x^{-1} = a\} \cdot \#\{y : 2 + y + y^{-1} = b\}.$$

Now, $2 + x + x^{-1} = a$ amounts to the quadratic equation $x^2 - (a - 2)x + 1 = 0$, which has $1 + \sigma(a(a - 4))$ solutions. Similarly, $2 + y + y^{-1} = b$ has $1 + \sigma(b(b - 4))$ solutions. Then

$$N(c) = \sum_{ab = c} \left(1 + \sigma\big(a(a - 4)\big) \right) \left(1 + \sigma\big(b(b - 4)\big) \right)$$

$$= p - 3 + \sum_{ab = c} \sigma\big(a(a - 4)b(b - 4)\big)$$

since $\sum_{a \in \mathbb{F}_p^*} \sigma\big(a(a-4)\big) = \sum_{b \in \mathbb{F}_p^*} \sigma\big(b(b-4)\big) = -1$. Moreover,

$$\sum_{ab=c} \sigma\big(a(a-4)b(b-4)\big) = \sigma(-c) \sum_{a \in \mathbb{F}_p^*} \sigma\big(a(a-4)(4a-c)\big).$$

In the latter sum, we include the index value $a = 0$, and we make the change of variable $a := -4a$. Summarizing, we obtain

$$N(c) = p - 3 + \sigma(c) \sum_{a \in \mathbb{F}_p} \sigma\left(a(a+1)\left(a + \frac{c}{16}\right)\right).$$

For $c = -16$, we obtain

$$N(-16) = p - 3 + \sigma(-16)\varphi_2(-1) = p - 3 + \sigma(-2)\varphi_2(1).$$

For $c \neq -16$, formula (125) yields

$$N(c) = p - 3 + \sigma(-2c(c+16))\varrho\left(\left(\frac{1}{2} \cdot \frac{c-16}{c+16}\right)^2\right).$$

In particular, we have

$$N(8) = p - 3 + \sigma(-6)\varrho\left(\frac{1}{36}\right) = p - 3 + \sigma(2)\varphi_2(1),$$

$$N(32) = p - 3 + \sigma(-3)\varrho\left(\frac{1}{36}\right) = p - 3 + \varphi_2(1).$$

Example 5.14. We count the number of solutions to the equation

$$x^3 + y^3 + 6xy = 4. \tag{136}$$

The change of variable $x := x - y$ turns the Equation (136) into

$$3(x-2)y^2 - 3x(x-2)y + x^3 - 4 = 0,$$

which we view as a quadratic equation in y. There are no solutions when $x = 2$. For each $x \neq 2$, there are $1 + \sigma(\Delta)$ solutions; here the discriminant Δ is given by

$$\Delta = 9x^2(x-2)^2 - 12(x-2)(x^3-4) = -3(x^2-4)(x^2+4x-8).$$

Thus, the number of solutions to the Equation (136) is

$$N = p - 1 + \sigma(-3) \sum_{x \in \mathbb{F}_p} \sigma\big((x^2 - 4)(x^2 + 4x - 8)\big).$$

The latter sum can be computed, thanks to Theorem 5.10, as follows:

$$\sum_{x \in \mathbb{F}_p} \sigma\big((x^2 - 4)(x^2 + 4x - 8)\big) = -1 + \sigma(-3)\varrho\left(\frac{1}{3}\right) = -1 + \sigma(-3)\psi_3(1).$$

We conclude that the number of solutions to the Equation (136) is

$$N = p - 1 - \sigma(-3) + \psi_3(1)$$

$$= \begin{cases} p - 2 + 2A_3(p) & \text{if } p \equiv 1 \bmod 3, \\ p & \text{if } p \equiv 2 \bmod 3. \end{cases}$$

Example 5.15. Consider again the generalized Markov cubic

$$x^2 + y^2 + z^2 = xyz + c, \tag{137}$$

where $c \in \mathbb{F}_p$. Recall that we have counted the number of solutions to (137) in Example 2.7. The set of solutions admits several obvious symmetries: coordinate permutations, and sign flips in two coordinates at a time. Somewhat less obvious are the Markov moves–the three involutions

$$m_1(x, y, z) = (yz - x, y, z),$$
$$m_2(x, y, z) = (x, xz - y, z),$$
$$m_3(x, y, z) = (x, y, xy - z).$$

These arise as follows: for fixed y and z, say, the generalized Markov cubic is a quadratic equation in x; the map m_1 interchanges the two solutions. This viewpoint explains why m_1, m_2, m_3 are also called Vieta involutions.

How many solutions to the generalized Markov cubic (137) are fixed by one of the Markov moves? Owing to the symmetry, we may consider either one of the moves–say m_1. A fixed point for m_1 has $x = yz/2$. Thus, solutions (x, y, z) to the generalized Markov cubic which are fixed by m_1 correspond to solutions (y, z) to the equation

$$y^2 + z^2 - \frac{y^2 z^2}{4} = c.$$

This can rearranged as

$$y^2(z^2 - 4) = 4(z^2 - c).$$

If $c = 4$, then one easily counts $4p - 4$ solutions.

If $c \neq 4$, then there are no solutions when $z^2 = 4$; when $z^2 \neq 4$, the above equation in y has $1 + \sigma\big((z^2 - 4)(z^2 - c)\big)$ solutions. Thus, the number of solutions is given by

$$\sum_{z^2 \neq 4} \big(1 + \sigma\big((z^2 - 4)(z^2 - c)\big)\big) = p - 2 + \sum_{z \in \mathbb{F}_p} \sigma\big((z^2 - 4)(z^2 - c)\big).$$

When $c = -4$, the latter sum equals $-1 + \varphi_2(-16) = -1 + \sigma(2)\varphi_2(1)$. When $c \neq -4$, it equals $-1 + \sigma(-(c+4))\varrho\big(4c/(c+4)^2\big)$, thanks to Theorem 5.10.

To summarize, the number of solutions to the generalized Markov cubic that are fixed by one of the Markov moves is given as follows:

$$\begin{cases} 4p - 4 & \text{if } c = 4, \\[2mm] p - 3 + \sigma(2)\varphi_2(1) & \text{if } c = -4, \\[4mm] p - 3 + \sigma(-(c+4))\varrho\left(\dfrac{4c}{(c+4)^2}\right) & \text{otherwise.} \end{cases}$$

The Cayley cubic, the case $c = 4$, stands out: the count is surprisingly large. The count in the case $c \neq \pm 4$ can be made explicit for certain values of the parameter: when $c = 2$, the outcome is $p - 3 + \sigma(-1)\varphi_2(1)$; when $c = 8$, the outcome is $p - 3 + \sigma(-2)\varphi_2(1)$; when $c = 0$, the case of the Markov cubic, the outcome is $p - 3 - \sigma(-1)$.

Quadratic character sums over some quartics II

Theorem 5.10 deals with certain reducible quartics. In the next result, we address a different type of quartics.

Theorem 5.16. *Let* $g(x) = (x^2 + ax + b)^2 + c(x + d)^2$, *where* $c \neq 0$. *Put*

$$r = d^2 - ad + b, \qquad s = a^2 - 4b + c$$

and assume that $r \neq 0$. *Then*

$$\sum_{x \in \mathbb{F}_p} \sigma\big(g(x)\big) = \begin{cases} -1 + \sigma(s)\varrho\left(\dfrac{-4cr}{s^2}\right) & \text{if } s \neq 0, \\[4mm] -1 + \varphi_2(r) & \text{if } s = 0. \end{cases} \tag{138}$$

The non-vanishing assumptions rule out some degenerate cases: if $c = 0$ then $g(x)$ is a square; if $r = 0$ then $(x+d)^2$ is a factor of $g(x)$.

Proof. In the given sum, we put aside the term corresponding to $x = -d$ for which we have $g(-d) = r^2$ and so $\sigma(g(-d)) = 1$. Then, using the trick already employed in the proofs of Theorems 5.3 and 5.10, we write

$$\sum_{x \in \mathbb{F}_p} \sigma\big(g(x)\big) = 1 + \sum_{x \neq -d} \sigma\big((x^2 + ax + b)^2 + c(x+d)^2\big)$$

$$= 1 + \sum_{x \neq -d} \sigma\left(\left(\frac{x^2 + ax + b}{x+d}\right)^2 + c\right)$$

$$= 1 + \sum_{y \in \mathbb{F}_p} \sigma(y^2 + c) \cdot \#\left\{x \in \mathbb{F}_p : y = \frac{x^2 + ax + b}{x+d}\right\}.$$

The quadratic equation $x^2 + ax + b = y(x+d)$ has $1 + \sigma(\Delta(y))$ solutions, the discriminant $\Delta(y)$ being given by

$$\Delta(y) = (a-y)^2 - 4(b - dy) = y^2 + 2(2d-a)y + (a^2 - 4b).$$

We note that $x = -d$ cannot be a solution of the above quadratic equation, as $d^2 - ad + b = r \neq 0$. Therefore,

$$\sum_{x \in \mathbb{F}_p} \sigma\big(g(x)\big) = 1 + \sum_{y \in \mathbb{F}_p} \sigma(y^2 + c)\big(1 + \sigma(\Delta(y))\big)$$

$$= \sum_{y \in \mathbb{F}_p} \sigma(y^2 + c)\sigma\big(y^2 + 2(2d-a)y + (a^2 - 4b)\big)$$

by using the familiar fact that $\sum_{y \in \mathbb{F}_p} \sigma(y^2 + c) = -1$ for $c \neq 0$. The latter sum can be evaluated by using Theorem 5.10. We have $\Delta_1 = -4c$, $\Delta_2 = 16r$, $B = 4s$, and so

$$\sum_{x \in \mathbb{F}_p} \sigma\big(g(x)\big) = -1 + \sigma(4s)\varrho\left(\frac{-4c \cdot 16r}{(4s)^2}\right) = -1 + \sigma(s)\varrho\left(\frac{-4cr}{s^2}\right)$$

when $s \neq 0$. For $s = 0$, we have

$$\sum_{x \in \mathbb{F}_p} \sigma\big(g(x)\big) = -1 + \varphi_2(-4c \cdot 16r) = -1 + \sigma(8)\varphi_2(-r) = -1 + \varphi_2(r).$$

The above simplification relies on (61) and (65). $\qquad\square$

Example 5.17. Let $p > 3$, and consider the character sum

$$\sum_{x \in \mathbb{F}_p} \sigma(x^4 - 18x^2 + 36x - 18). \tag{139}$$

The quartic argument does not admit an integral factorization into two quadratics, so Theorem 5.10 is not applicable. But we can use Theorem 5.16 above, since

$$x^4 - 18x^2 + 36x - 18 = x^4 - 18(x - 1)^2.$$

Here, $a = b = 0$, $c = -18$, $d = -1$ so $r = 1$ and $s = -18$. Noting that $c, r, s \neq 0$, we see that the given sum evaluates as

$$-1 + \sigma(-18)\varrho\left(\frac{2}{9}\right) = -1 + \sigma(-3)\varphi_2(1).$$

The next example is of a more general nature. It addresses the case of quadratic character sums whose argument is a palindromic quartic.

Example 5.18. Let $a, b \in \mathbb{F}_p$, and consider the quadratic character sum

$$\sum_{x \in \mathbb{F}_p} \sigma(x^4 + ax^3 + bx^2 + ax + 1). \tag{140}$$

Denote the sum by Σ. Aiming for an application of Theorem 5.16, we write

$$x^4 + ax^3 + bx^2 + ax + 1 = (x^2 + Ax + 1)^2 + Bx^2$$

with $A = a/2$ and $B = b - 2 - a^2/4$.

When $b - 2 - a^2/4 = 0$ the evaluation is straightforward:

$$\Sigma = \sum_{x \in \mathbb{F}_p} \sigma\big((x^2 + Ax + 1)^2\big)$$

$$= p - \#\{x \in \mathbb{F}_p : x^2 + Ax + 1 = 0\}$$

$$= p - \big(1 + \sigma(A^2 - 4)\big) = p - 1 - \sigma(a^2 - 16).$$

When $b - 2 - a^2/4 \neq 0$ we may use Theorem 5.16 with $r = 1$ and $s = A^2 - 4 + B = b - 6$:

$$\Sigma = \begin{cases} -1 + \sigma(b - 6)\varrho\left(\dfrac{a^2 - 4(b - 2)}{(b - 6)^2}\right) & \text{if } b \neq 6, \\[4mm] -1 + \varphi_2(1) & \text{if } b = 6. \end{cases}$$

The method used in the proof of Theorem 5.16 can be modified to handle quartics of the form $(x^2 + ax + b)^2 + c$. A similar, though slightly easier, argument leads to a ϱ-sum formula analogous to (138). We illustrate the approach in the following example.

Example 5.19. Let $c \in \mathbb{F}_p$, and consider the quadratic character sum

$$\sum_{x \in \mathbb{F}_p} \sigma\big((x + 2)(x^3 - 1) + c\big). \tag{141}$$

We write the quartic argument as follows:

$$(x + 2)(x^3 - 1) + c = x^4 + 2x^3 - x + c - 2$$
$$= \left(x^2 + x - \frac{1}{2}\right)^2 + C, \quad C := c - \frac{9}{4}.$$

Let the sum in (141) be denoted, as usual, by Σ. Then

$$\Sigma = \sum_{x \in \mathbb{F}_p} \sigma\left(\left(x^2 + x - \frac{1}{2}\right)^2 + C\right)$$
$$= \sum_{y \in \mathbb{F}_p} \sigma(y^2 + C) \cdot \# \left\{x \in \mathbb{F}_p : y = x^2 + x - \frac{1}{2}\right\}$$
$$= \sum_{y \in \mathbb{F}_p} \sigma(y^2 + C)\big(1 + \sigma(4y + 3)\big).$$

When $C = 0$, we simply have

$$\Sigma = \sum_{y \neq 0} \big(1 + \sigma(4y + 3)\big) = p - 1 - \sigma(3).$$

Assume $C \neq 0$ in what follows. Then $\sum_{y \in \mathbb{F}_p} \sigma(y^2 + C) = -1$, and so

$$\Sigma = -1 + \sum_{y \in \mathbb{F}_p} \sigma\big((4y + 3)(y^2 + C)\big).$$

The change of variable $y := y - 3/4$ leads to

$$\Sigma = -1 + \sum_{y \in \mathbb{F}_p} \sigma\left(y\left(y^2 - \frac{3}{2}y + \frac{9}{16} + C\right)\right)$$
$$= -1 + \sigma\left(-\frac{3}{2}\right)\varrho\left(\frac{9/16 + C}{9/4}\right)$$

as long as $p > 3$. Using the twisted symmetry of ϱ-sums, and reverting to the original parameter c, we conclude that

$$\Sigma = -1 + \sigma(3)\varrho\left(\frac{-4C}{9}\right) = -1 + \sigma(3)\varrho\left(\frac{9 - 4c}{9}\right).$$

The evaluation of Σ can be made explicit for a handful of values such as $c = 3/2$ and $c = 7/4$.

Quadratic character sums over some quintics

The following result is a closed-form formula, in terms of ϱ-sums, for quadratic character sums whose argument is a palindromic quintic.

Theorem 5.20. *Let*

$$S(a, b) = \sum_{x \in \mathbb{F}_p} \sigma(x^5 + ax^4 + bx^3 + ax^2 + x). \tag{142}$$

If $a \neq \pm 4$ then

$$S(a, b) = \sigma(a + 4)\varrho\left(\frac{b + 2a + 2}{(a + 4)^2}\right) + \sigma(a - 4)\varrho\left(\frac{b - 2a + 2}{(a - 4)^2}\right). \tag{143}$$

Proof. We start by observing that the change of variable $x := -x$ yields the relation $S(-a, b) = \sigma(-1)S(a, b)$. This twisted symmetry is visible in formula (143) as well, and we will use it below.

The bulk of the proof is concerned with proving (143) when $b \neq -2a-2$. Once this is done, we may round off the proof as follows. If $b = -2a - 2$ but $b \neq 2a - 2$, we invoke the twisted symmetry $S(a, b) = \sigma(-1)S(-a, b)$; now formula (143) can be used for $S(-a, b)$, since $b \neq 2a - 2$, and (143) for $S(a, b)$ follows. Finally, if $b = -2a - 2$ and $b = 2a - 2$, then $a = 0$ and $b = -2$; in this case (143) claims that

$$S(0, -2) = \sigma(4)\varrho(0) + \sigma(-4)\varrho(0) = -\sigma(1) - \sigma(-1).$$

This can be easily checked:

$$\sum_{x \in \mathbb{F}_p} \sigma(x^5 - 2x^3 + x) = \sum_{x \in \mathbb{F}_p} \sigma\left(x(x^2 - 1)^2\right) = \sum_{x \neq \pm 1} \sigma(x)$$

$$= -\sigma(1) - \sigma(-1).$$

We turn to the main case of interest, when $b \neq -2a - 2$. We write

$$S(a,b) = \sum_{x \in \mathbb{F}_p} \sigma\big(x(x^4 + ax^3 + bx^2 + ax + 1)\big)$$

and we put the quartic factor in the following form:

$$x^4 + ax^3 + bx^2 + ax + 1 = \alpha(x+1)^4 + \beta(x^2 - 1)^2 + \gamma(x-1)^4.$$

Upon expanding the right-hand side, we see that the coefficients α, β, γ are determined by the conditions

$$\alpha + \beta + \gamma = 1, \quad 4(\alpha - \gamma) = a, \quad 6(\alpha + \gamma) - 2\beta = b.$$

The explicit solution is

$$\alpha = \frac{b + 2a + 2}{16}, \qquad \beta = \frac{6 - b}{8}, \qquad \gamma = \frac{b - 2a + 2}{16}. \qquad (144)$$

The meaning of our assumption, $b \neq -2a - 2$, is now clear: it amounts to $\alpha \neq 0$. We then write

$$S(a,b) = \sum_{x \in \mathbb{F}_p} \sigma\Big(x\big(\alpha(x+1)^4 + \beta(x^2 - 1)^2 + \gamma(x-1)^4\big)\Big)$$

$$= \sigma(\alpha) + \sum_{x \neq 1} \sigma\Big(x\big(\alpha(x+1)^4 + \beta(x^2 - 1)^2 + \gamma(x-1)^4\big)\Big)$$

$$= \sigma(\alpha) + \sum_{x \neq 1} \sigma\left(x\left(\alpha\left(\frac{x+1}{x-1}\right)^4 + \beta\left(\frac{x+1}{x-1}\right)^2 + \gamma\right)\right).$$

The map $x \mapsto \frac{x+1}{x-1}$ is a bijection from $\mathbb{F}_p \backslash \{1\}$ to itself; in fact, it is an involution. Therefore,

$$S(a,b) = \sigma(\alpha) + \sum_{x \neq 1} \sigma\left(\frac{x+1}{x-1} \cdot (\alpha x^4 + \beta x^2 + \gamma)\right)$$

$$= \sigma(\alpha) + \sum_{x \neq 1} \sigma\big((x^2 - 1)(\alpha x^4 + \beta x^2 + \gamma)\big)$$

$$= \sigma(\alpha) + \sum_{x \in \mathbb{F}_p} \sigma\big((x^2 - 1)(\alpha x^4 + \beta x^2 + \gamma)\big).$$

By the downgrading formula, we can now write

$$S(a,b) = \sigma(\alpha) + S_3 + S_4,$$

where

$$S_3 = \sum_{x \in \mathbb{F}_p} \sigma\big((x-1)(\alpha x^2 + \beta x + \gamma)\big) = \sum_{x \in \mathbb{F}_p} \sigma\big(x(\alpha x^2 + (2\alpha + \beta)x + 1)\big),$$

$$S_4 = \sum_{x \in \mathbb{F}_p} \sigma\big(x(x-1)(\alpha x^2 + \beta x + \gamma)\big) = \sum_{x \in \mathbb{F}_p} \sigma\big((x^2 - x)(\alpha x^2 + \beta x + \gamma)\big).$$

To clarify, the rewriting of the sum S_3 stems from the change of variable $x := x + 1$; the relation $\alpha + \beta + \gamma = 1$ is used along the way.

We turn to evaluating S_3 and S_4. We note that the polynomial argument in S_3 is a reducible cubic while that in S_4 is a product of two quadratics; for each one we have computational techniques in place. To begin with, by using (117), we have

$$S_3 = \sigma(\alpha) \sum_{x \in \mathbb{F}_p} \sigma\left(x\left(x^2 + \frac{2\alpha + \beta}{\alpha}x + \frac{1}{\alpha}\right)\right)$$

$$= \sigma(\alpha)\sigma\left(\frac{2\alpha + \beta}{\alpha}\right)\varrho\left(\frac{\alpha}{(2\alpha + \beta)^2}\right) = \sigma(2\alpha + \beta)\varrho\left(\frac{\alpha}{(2\alpha + \beta)^2}\right).$$

We note that $2\alpha + \beta = (a + 4)/4 \neq 0$, per our hypothesis that $a \neq -4$.

As for the sum S_4, we write

$$S_4 = \sigma(\alpha) \sum_{t \in \mathbb{F}_p} \sigma\left((t^2 - t)\left(t^2 + \frac{\beta}{\alpha}t + \frac{\gamma}{\alpha}\right)\right)$$

and we resort to Theorem 5.10. The relevant parameters are

$$\Delta_1 = 1, \qquad \Delta_2 = \frac{\beta^2 - 4\alpha\gamma}{\alpha^2}, \qquad B = \frac{2(2\gamma + \beta)}{\alpha}.$$

We now have $2\gamma + \beta = (4 - a)/4 \neq 0$, in view of the hypothesis that $a \neq 4$. Thus $B \neq 0$, and we obtain

$$S_4 = -\sigma(\alpha) + \sigma(\alpha)\sigma\left(\frac{2(2\gamma + \beta)}{\alpha}\right)\varrho\left(\frac{\beta^2 - 4\alpha\gamma}{4(2\gamma + \beta)^2}\right)$$

$$= -\sigma(\alpha) + \sigma(2(2\gamma + \beta))\varrho\left(\frac{\beta^2 - 4\alpha\gamma}{4(2\gamma + \beta)^2}\right).$$

The latter ϱ-sum can be simplified through the twisted symmetry, Theorem 5.3, as follows:

$$\varrho\left(\frac{\beta^2 - 4\alpha\gamma}{4(2\gamma + \beta)^2}\right) = \sigma(-2)\varrho\left(\frac{1}{4} - \frac{\beta^2 - 4\alpha\gamma}{4(2\gamma + \beta)^2}\right).$$

The relation $\alpha + \beta + \gamma = 1$ yields

$$\frac{1}{4} - \frac{\beta^2 - 4\alpha\gamma}{4(2\gamma + \beta)^2} = \frac{\gamma}{(2\gamma + \beta)^2}.$$

Therefore,

$$S_4 = -\sigma(\alpha) + \sigma(-(2\gamma + \beta))\varrho\left(\frac{\gamma}{(2\gamma + \beta)^2}\right).$$

The end result of our computations is that

$$S(a,b) = \sigma(2\alpha + \beta)\varrho\left(\frac{\alpha}{(2\alpha + \beta)^2}\right) + \sigma(-(2\gamma + \beta))\varrho\left(\frac{\gamma}{(2\gamma + \beta)^2}\right).$$

The last step, and the simplest, is to return to the original parameters a and b. We replace α and β using (144); we also recall that $2\alpha + \beta = (a+4)/4$ and $2\gamma + \beta = -(a - 4)/4$. Formula (143) quickly follows. $\qquad\square$

Example 5.21. For $p > 3$, we have

$$\sum_{x \in \mathbb{F}_p} \sigma(x^5 + 28x^4 + 70x^3 + 28x^2 + x) = \sigma(32)\varrho\left(\frac{1}{8}\right) + \sigma(24)\varrho\left(\frac{1}{36}\right)$$

$$= \sigma(2)\varrho\left(\frac{1}{8}\right) + \sigma(-2)\varphi_2(1).$$

Here is a notable instance of Theorem 5.20.

Corollary 5.22. *We have*

$$\sum_{x \in \mathbb{F}_p} \sigma(x^5 + bx^3 + x) = \begin{cases} 2\varrho\left(\dfrac{b + 2}{16}\right) & \text{if } p \equiv 1 \bmod 4, \\ 0 & \text{if } p \equiv 3 \bmod 4. \end{cases} \tag{145}$$

An unexpected outcome of Corollary 5.22 is the following evaluation of a ϱ-sum.

Corollary 5.23. *Assume $p \equiv 1 \bmod 8$. Then*

$$\varrho\left(\frac{1}{8}\right) = (-1)^{(p-1)/8} \cdot 2A_4(p). \tag{146}$$

Proof. The case $b = 0$ of Corollary 5.22 amounts to

$$\varphi_4(1) = 2\varrho\left(\frac{1}{8}\right),$$

whenever $p \equiv 1 \bmod 4$. This formula is not terribly interesting in the case $p \equiv 5 \bmod 8$, since both sides vanish. When $p \equiv 1 \bmod 8$, we may use the computation of the Jacobsthal sum $\varphi_4(1)$ given in Theorem 3.18(iii). $\qquad\square$

Exercises

Exercise 5.24. Show that

$$\sum_{c \in \mathbb{F}_p} \varrho(c) = 0, \qquad \sum_{c \in \mathbb{F}_p} \varrho(c)^2 = p(p - 2 - \sigma(-1)).$$

Exercise 5.25. Let $a \in \mathbb{F}_p$. Evaluate

$$\sum_{c \in \mathbb{F}_p} \varrho(c)\varrho(ac), \qquad \sum_{c \in \mathbb{F}_p} \varrho(c)\varrho(a + c).$$

Exercise 5.26. Let $c \in \mathbb{F}_p$, with $c \neq 0, 1/4$. Show that $\varrho(c)$ modulo 4 is given as follows:

$$\varrho(c) \equiv \begin{cases} \sigma(-1) - 1 & \text{if } \sigma(c) = \sigma(1 - 4c) = -1, \\ \sigma(-1) + 1 & \text{otherwise.} \end{cases}$$

Exercise 5.27. Let $a, c, d \in \mathbb{F}_p$. Show that

$$\sum_{x \in \mathbb{F}_p} \sigma(x)\sigma\left(a^2 + acx + dx^2\right) = \sum_{x \in \mathbb{F}_p} \sigma(-x)\sigma\left(a^2 + 2acx + (c^2 - 4d)x^2\right).$$

Exercise 5.28. Let $p > 3$. Evaluate as explicitly as possible the following sums of Legendre symbols:

$$\sum_{x=0}^{p-1} \left(\frac{x^4 - 6x^2 - 16x + 41}{p}\right),$$

$$\sum_{x=0}^{p-1}\left(\frac{9x^5 + 36x^4 + 38x^3 + 36x^2 + 9x}{p}\right),$$

$$\sum_{x=0}^{p-1}\left(\frac{x^6 - 15x^4 + 15x^2 - 1}{p}\right).$$

Exercise 5.29. (i) Let $b, c \in \mathbb{F}_p$ with $c \neq 0$. Show that

$$\sum_{x\in\mathbb{F}_p} \sigma(x^5 + bcx^3 + c^2x) = \begin{cases} \pm 2\varrho\left(\dfrac{b+2}{16}\right) & \text{if } p \equiv 1 \bmod 4,\ \sigma(c) = 1, \\ 0 & \text{otherwise.} \end{cases}$$

(ii) Let $p > 3$ and let $a \in \mathbb{F}_p^*$. Show that

$$\sum_{x\in\mathbb{F}_p} \sigma(x^5 + 10ax^3 + 9a^2x) = \begin{cases} \pm 4A_3(p) & \text{if } p \equiv 1 \bmod 12,\ \sigma(a) = 1, \\ 0 & \text{otherwise.} \end{cases}$$

Give a formula for the sign in the case $a = -1$.

Exercise 5.30. Evaluate as explicitly as possible the sum of Legendre symbols

$$\sum_{x=0}^{p-1}\left(\frac{x^7 + ax^5 + ax^3 + x}{p}\right),$$

for the two parameter values, $a = 7$ and $a = -33$.

Exercise 5.31. Let $p > 3$. Count the number of solutions to the equation

$$(x + y + z)^3 = 54xyz.$$

Exercise 5.32. Let $p > 3$. Count the number of solutions to the equation

$$(x + y + z + t + 1)^2 = 32xyzt.$$

Exercise 5.33. For $b \in \mathbb{F}_p^*$, we let $N(b)$ denote the number of solutions to the equation

$$x + \frac{1}{x} + y + \frac{1}{y} = 4b.$$

Show that

$$N\left(\frac{1}{b}\right) = N(b).$$

Notes

- The technique used in the proofs of Theorems 5.3, 5.10, and 5.16 can already be discerned in Jacobsthal's work. Subsequently, a passing mention of this technique can be found in a paper of Davenport (*On certain exponential sums*, J. Reine Angew. Math. 1933). But it was Williams who, in a number of papers (*Finite transformation formulae involving the Legendre symbol*, Pacific J. Math. 1970; *Note on Brewer's character sum*, Proc. Amer. Math. Soc. 1978; *Evaluation of character sums connected with elliptic curves*, Proc. Amer. Math. Soc. 1979), codified and used the technique in a systematic way. Theorems 5.10 and 5.16 are due to Williams, though framed herein as formulas in terms of ϱ-sums.

  In particular, Jacobsthal obtained the following identity in his thesis (*Anwendungen einer Formel aus der Theorie der quadratischen Reste*, Dissertation, Berlin, 1906): for $a, b \in \mathbb{F}_p$ it holds that

$$\sum_{x \in \mathbb{F}_p} \sigma(x)\sigma(x^2 + ax + b) = \sum_{x \in \mathbb{F}_p} \sigma(x + a)\sigma(x^2 - 4b). \tag{147}$$

  This identity has been noted, independently, by Parnami, Agrawal, and Rajwade (*Some identities involving character sums and their applications*, J. Indian Math. Soc. 1989). From the viewpoint of ϱ-sums, (147) is essentially equivalent to the twisted symmetry of Theorem 5.3. The equivalence is mediated by formula (117), as long as $a \neq 0$.

- The Brewer sums form a family of quadratic character sums with polynomial arguments, the first non-trivial member of which is the sum $\varrho(1/8)$. They were introduced by Brewer (*On certain character sums*, Trans. Amer. Math. Soc. 1961). Brewer computed the sum herein denoted $\varrho(1/8)$, by using mod p congruences for binomial coefficients. Later on, Whiteman (*A theorem of Brewer on character sums*, Duke Math. J. 1963) offered a more direct approach. The outcome reads as follows:

$$\varrho\left(\frac{1}{8}\right) = \begin{cases} 0 & \text{if } p \equiv 5, 7 \bmod 8, \\ (-1)^k \cdot 2A & \text{if } p \equiv 1, 3 \bmod 8. \end{cases}$$

Here, A is determined by the sum-of-squares representation $p = A^2 + 2B^2$ and the modular condition $A \equiv -1 \bmod 4$, while $k = (p-1)/8$ when $p \equiv 1 \bmod 8$, respectively, $k = (p-3)/8$ when $p \equiv 3 \bmod 8$. We have handled the vanishing case, $p \equiv 5, 7 \bmod 8$, in Corollary 5.6, and the case

$p \equiv 1 \bmod 8$ in Corollary 5.23. The most difficult case is $p \equiv 3 \bmod 8$. With the above evaluation of $\varrho(1/8)$ at hand, one can make Example 5.21 fully explicit.

We have already mentioned that Hashimoto, Long, and Yang (*Jacobsthal identity for* $\mathbb{Q}(\sqrt{-2})$, Forum Math. 2012) have obtained a representation $p = A^2 + 2B^2$ for $p \equiv 3 \bmod 8$, in which A and B are explicitly given by certain quadratic character sums. In view of the above computation, it should not be surprising that they take $A = \frac{1}{2}\varrho(1/8)$.

- Poulakis (*Évaluation d'une somme cubique de caractères*, J. Number Theory 1987) determined $\varrho(-1/12)$ in a sophisticated way, by means of elliptic curves.

- Rajwade (*The Diophantine equation $y^2 = x(x^2+21Dx+112D^2)$ and the conjectures of Birch and Swinnerton-Dyer*, J. Austral. Math. Soc. Ser. A 1977; *On a conjecture of Williams*, Bull. Soc. Math. Belg. Sér. B 1984) has evaluated

$$\sum_{x \in \mathbb{F}_p} \sigma(x^3 + 21x^2 + 112x) = \begin{cases} 2A & \text{if } p \equiv 1, 2, 4 \bmod 7, \\ 0 & \text{if } p \equiv 3, 5, 6 \bmod 7, \end{cases}$$

where A is determined by writing $p = A^2 + 7B^2$ and choosing the sign of A so that A is a quadratic non-residue mod 7; that is, $A \equiv -1, 3, -2 \bmod 7$ respectively as $p \equiv 1, 2, 4 \bmod 7$.

In the language of ϱ-sums, this amounts to the evaluation of $\varrho(16/63)$ or, in view of the twisted symmetry, to the evaluation of $\varrho(-1/252)$. Rajwade's result is the only explicit evaluation of a ϱ-sum, beside the ones given in this section, that we are aware of. However, Rajwade's method is non-elementary.

- The double sum appearing in Example 5.9 can be computed as follows:

$$\sum_{x,y \in \mathbb{F}_p} \sigma\big(xy(x+1)(y+1)(x+y)\big) = \begin{cases} \sigma(-1)p & \text{if } p \equiv 5, 7 \bmod 8, \\ \sigma(-1)(4A^2 - p) & \text{if } p \equiv 1, 3 \bmod 8 \end{cases}$$

where the latter case involves the representation $p = A^2 + 2B^2$. This formula was conjectured by Evans, Pulham, and Sheehan (*On the number of complete subgraphs contained in certain graphs*, J. Combin. Theory Ser. B 1981). Emma Lehmer (unpublished) promptly proved the $p \equiv 5, 7$ mod 8 case in an elementary way, exploiting once again the xy-to-uv transformation used in solving Lehmers' equation. Example 5.9 also handles the $p \equiv 5, 7$ mod 8 case by an elementary argument. The evaluation in both cases was settled by Greene and Stanton (*A character sum evaluation and Gaussian hypergeometric series*, J. Number Theory 1986), by means of hypergeometric functions.

The general double sum

$$I(t) = \sum_{x,y\in\mathbb{F}_p} \sigma\big(xy(x+1)(y+1)(x+ty)\big) \qquad (t \in \mathbb{F}_p^*)$$

was investigated by Ono (*Values of Gaussian hypergeometric series*, Trans. Amer. Math. Soc. 1998), still exploiting the technology of hypergeometric functions. Ono evaluated $I(t)$ for $t \in \{1, -1, -4, 8, -64\}$; the values for $t \in \{-1/4, 1/8, -1/64\}$ follow, in view of the obvious relation $I(1/t) = \sigma(t)I(t)$. The evaluations involve, just like the case $t = 1$ discussed above, representations of p as sums of squares. We note that an elementary approach to the case $t = -1$ appears in Exercise 4.21. The evaluation of $I(t)$ for $t = -1$, in fact a generalization thereof, was first obtained by Evans (*Identities for products of Gauss sums over finite fields*, Enseign. Math. 1981).

Later on, Ahlgren, Ono, and Penniston (*Zeta functions of an infinite family of K3 surfaces*, Amer. J. Math. 2002) obtained an interesting formula for $I(t)$, as long as $t \neq -1$. Their elementary, though elaborate, approach relies on Jacobi sums. The formula in question can be expressed in the language ϱ-sums as follows:

$$\sigma(t+1)I(t) = \varrho\left(\frac{1}{4(t+1)}\right)^2 - p.$$

We immediately see that, for the parameters $t \in \{1, -4, 8, -64\}$, the ϱ-sums appearing on the right-hand side are $\varrho(1/8)$, $\varrho(-1/12)$, $\varrho(1/36)$, $\varrho(-1/252)$. All these have known evaluations.

- Example 5.15 is inspired by counting results considered by Cerbu, Gunther, Magee, and Peilen (*The cycle structure of a Markoff automorphism over finite fields*, J. Number Theory 2020) and, very recently, by de Courcy-Ireland (*Non-planarity of Markoff graphs* mod p, Comment. Math. Helv. 2024).

- Corollary 5.22 recovers a result due to Leprévost and Morain (*Revêtements de courbes elliptiques à multiplication complexe par des courbes hyperelliptiques et sommes de caractères*, J. Number Theory 1997).

- Exercise 5.27 is a result of Krachun, Petrov, Zhi-Wei Sun, and Vsemirnov (*On some determinants involving Jacobi symbols*, Finite Fields Appl. 2020), though it is not hard to see that it is equivalent to Jacobsthal's identity (147). Exercise 5.29 is inspired by, and sharpens results from the same reference. Our viewpoint is that part (i) of Exercise 5.29 expands on Corollary 5.22. The methods of Krachun, Petrov, Sun, and Vsemirnov are very different.

- Exercise 5.32 is a variation on a result due to Carlitz (*The number of solutions of some equations in a finite field*, Portugal. Math. 1954).

- Exercise 5.33 is based on a question raised by Emmanuel Kowalski (https://blogs.ethz.ch/kowalski/2012/06/29/a-bijective-challenge). For further context, and an alternate approach, see the interesting report by Budzinski and Lucchini Arteche (*Un résultat de recherche obtenu grâce à des lycéens*, Gaz. Math. 2017).

Solutions to Exercises

Chapter 1

Exercise 1.30. If a prime p is of the form $p = A^2 + B^2$ for some integers A, B, then $z = A/B \in \mathbb{F}_p$ satisfies $z^2 = -1$. For $p = 409$, we have $p = 20^2 + 3^2$, so $z = 20/3 = 7 - 1/3$ in $\mathbb{F}_p$. The multiplicative inverse of 3 is -136. Thus, a square root of -1 modulo $p = 409$ is $z = 143$.

If a prime p is of the form $p = C^2 - 2D^2$ for some integers C, D, then $w = C/D \in \mathbb{F}_p$ satisfies $w^2 = 2$. For $p = 409$, we have $p = 21^2 - 2 \cdot 4^2$, so $w = 21/4 = 5 + 1/4$ in $\mathbb{F}_p$. The multiplicative inverse of 4 is -102. Thus, a square root of 2 modulo $p = 409$ is $w = 312$.

Finally, we apply Theorem 1.25. We represent $p = 11^2 + 2 \cdot 12^2$. Since $11 \equiv 3 \bmod 8$, we conclude that 2 is not a fourth power modulo $p = 409$.

Exercise 1.31. Thinking of a matrix $\left(\begin{smallmatrix} a & b \\ c & d \end{smallmatrix}\right) \in \mathrm{SL}_2(\mathbb{F}_p)$, we have to count the number of solutions $(a, b, c, d) \in \mathbb{F}_p^4$ to the system $ad - bc = 1$, $a + d = t$. This system amounts, quite simply, to the single equation $-bc = a^2 - ta + 1$. There are $p - 1$ solutions when $a^2 - ta + 1 \neq 0$, respectively, $2p - 1$ solutions when $a^2 - ta + 1 = 0$. The total number of solutions is

$$p(p-1) + p \cdot \#\{a \in \mathbb{F}_p : a^2 - ta + 1 = 0\} = p(p-1) + p(1 + \sigma(t^2 - 4))$$

$$= p^2 + \sigma(t^2 - 4) \cdot p.$$

Since we are partitioning $\mathrm{SL}_2(\mathbb{F}_p)$ according to trace values, and $\mathrm{SL}_2(\mathbb{F}_p)$ has size $p^3 - p$, the identity $\sum_{t \in \mathbb{F}_p} (p^2 + \sigma(t^2 - 4) \cdot p) = p^3 - p$ should hold. And it does: $\sum_{t \in \mathbb{F}_p} \sigma(t^2 - 4) = -1$ by the forthcoming formula (34).

Exercise 1.32. (i) Entry-wise, we have

$$Q_{ab}^{\mathsf{T}} = \left(\frac{b-a}{p}\right) = \left(\frac{-1}{p}\right)\left(\frac{a-b}{p}\right) = \left(\frac{-1}{p}\right) Q_{ab}.$$

(ii) We compute the entries of Q^2 by using Theorem 1.11:

$$Q_{ab}^2 = \sum_{c=1}^{p} Q_{ac}\, Q_{cb} = \sum_{c=1}^{p} \left(\frac{a-c}{p}\right)\left(\frac{c-b}{p}\right) = \begin{cases} -(-1/p) & \text{if } a \neq b, \\ (-1/p)(p-1) & \text{if } a = b. \end{cases}$$

In matrix form, $Q^2 = (-1/p) \cdot (pI_p - J_p)$.

Exercise 1.33. (i) Clearly, the entries of H are 1 or -1. As $p \equiv 3 \bmod 4$, we have $Q^{\mathsf{T}} = -Q$ and $Q^2 = -(pI_p - J_p)$. Using these two facts we compute

$$HH^{\mathsf{T}} = \begin{pmatrix} 1 & 1 & \cdots & 1 \\ 1 & & & \\ \vdots & & Q - I_p & \\ 1 & & & \end{pmatrix} \begin{pmatrix} 1 & 1 & \cdots & 1 \\ 1 & & & \\ \vdots & & -Q - I_p & \\ 1 & & & \end{pmatrix}$$

$$= \begin{pmatrix} p+1 & 0 & \cdots & 0 \\ 0 & & & \\ \vdots & & J_p - (Q^2 - I_p) & \\ 0 & & & \end{pmatrix} = (p+1)I_{p+1}.$$

(ii) The matrices M_+ and M_- have entries 1 or -1. As M_+ and M_- are different perturbations of the same underlying matrix by multiples of the identity matrix I_{p+1}, they commute: $M_+ M_- = M_- M_+$. Furthermore, M_+ and M_- are symmetric since Q is; the latter owes to the hypothesis that $p \equiv 1 \bmod 4$.

The matrix

$$\tilde{H} = \begin{pmatrix} M_+ & M_- \\ M_- & -M_+ \end{pmatrix},$$

has entries 1 or -1. We compute

$$\tilde{H}\tilde{H}^{\mathsf{T}} = \begin{pmatrix} M_+ & M_- \\ M_- & -M_+ \end{pmatrix}\begin{pmatrix} M_+ & M_- \\ M_- & -M_+ \end{pmatrix} = \begin{pmatrix} M_+^2 + M_-^2 & 0_{p+1} \\ 0_{p+1} & M_+^2 + M_-^2 \end{pmatrix}.$$

Checking that $\tilde{H}\tilde{H}^{\mathsf{T}} = 2(p+1)I_{2(p+1)}$ boils down to

$$M_+^2 + M_-^2 = 2(p+1)I_{p+1}. \tag{$*$}$$

The square of each M matrix can be evaluated as follows:

$$M_\pm^2 = \begin{pmatrix} \pm 1 & 1 & \cdots & 1 \\ 1 & & & \\ \vdots & & Q \pm I_p & \\ 1 & & & \end{pmatrix} \begin{pmatrix} \pm 1 & 1 & \cdots & 1 \\ 1 & & & \\ \vdots & & Q \pm I_p & \\ 1 & & & \end{pmatrix}$$

$$= \begin{pmatrix} p+1 & \pm 1 & \cdots & \pm 1 \\ \pm 1 & & & \\ \vdots & & J_p + (Q \pm I_p)^2 & \\ \pm 1 & & & \end{pmatrix}.$$

Now, $Q^2 = pI_p - J_p$, as $p \equiv 1 \bmod 4$. Therefore, $J_p + (Q \pm I_p)^2 = (p+1)I_p \pm 2Q$, and $(*)$ follows.

Exercise 1.34. The change of variables $x := x + b/4$, $y := y + b/4$, $z := z + b/4$, $t := t + b/4$ brings us to the system

$$\begin{cases} x^2 + y^2 + z^2 + t^2 = a - b^2/4 \\ x + y + z + t = 0 \end{cases}$$

which, upon eliminating the variable t, turns into the single equation

$$x^2 + y^2 + z^2 + (x + y + z)^2 = a - b^2/4.$$

Equivalently,

$$(x + y)^2 + (y + z)^2 + (z + x)^2 = a - b^2/4.$$

The second change of variables, $Z := x+y$, $X := y+z$, $Y := z+x$, turns the above equation into $X^2 + Y^2 + Z^2 = a - b^2/4$. According to Example 1.20, the latter equation has $p^2 + \sigma(b^2 - 4a)p$ solutions.

Exercise 1.35. Let S_c denote the sphere described by $x_1^2 + \cdots + x_n^2 = c$ in $\mathbb{F}_p^n$. With an eye towards the uniform case of Theorem 1.19, we break the argument into two cases.

Case 1: $c \neq 0$, or n is odd. The hyperplane $x_n = ax_1$ meets the sphere S_c in as many points as there are on the conic

$$(1 + a^2)x_1^2 + x_2^2 + \cdots + x_{n-1}^2 = c$$

in $\mathbb{F}_p^{n-1}$. As long as $1 + a^2 \neq 0$, the latter point count depends on $\sigma(1 + a^2)$. As $\sigma(1 + a^2) = 1$ for $a = 0$, slice uniformity will fail if we can pick a scalar $a \in \mathbb{F}_p$ so that $\sigma(1 + a^2) = -1$. Let $b \in \mathbb{F}_p^*$ be a non-square. The quadratic equation $1 + a^2 = by^2$ has $p - \sigma(b) = p + 1$ solutions $(a, y) \in \mathbb{F}_p^2$; at most 2 solutions have $y = 0$, so there exist solutions (a, y) with $y \neq 0$. Then $\sigma(1 + a^2) = \sigma(b) = -1$, as desired.

Case 2: $c = 0$, and n is even. Consider first the lowest possible dimension, $n = 2$. Then S_0 is simply $x_1^2 + x_2^2 = 0$, and a line through the origin in $\mathbb{F}_p^2$ has the form $x_2 = ax_1$, up to a coordinate switch. The intersection is described by $(1 + a^2)x_1^2 = 0$. Slice uniformity amounts to ruling out the possibility that $1 + a^2 = 0$; that is, requiring $p \equiv 3 \bmod 4$.

Consider now the high-dimensional case $n \geq 4$. The hyperplane $x_n = x_1$ meets the sphere S_0 in as many points as there are on the conic

$$2x_1^2 + x_2^2 + \cdots + x_{n-1}^2 = 0$$

in $\mathbb{F}_p^{n-1}$; that is, p^{n-2} points. On the other hand, the hyperplane $x_n = ax_1 + bx_2$ meets the sphere S_0 in as many points as there are on the conic

$$(1 + a^2)x_1^2 + (1 + b^2)x_2^2 + 2abx_1x_2 + x_3^2 + \cdots + x_{n-1}^2 = 0$$

in $\mathbb{F}_p^{n-1}$. Choose $a, b \in \mathbb{F}_p$ so that $a^2 + b^2 = -1$. Then we are, in fact, looking at the conic

$$-(bx_1 - ax_2)^2 + x_3^2 + \cdots + x_{n-1}^2 = 0$$

in $\mathbb{F}_p^{n-1}$. The equation $-y^2 + x_3^2 + \cdots + x_{n-1}^2 = 0$ has $p^{n-3} + \sigma(-1)^{n/2}(p - 1)p^{(n-4)/2}$ solutions; and for each y, there are p solutions to $y = bx_1 - ax_2$. Thus, we are obtaining $p^{n-2} + \sigma(-1)^{n/2}(p-1)p^{(n-2)/2}$ intersection points–certainly different from p^{n-2}.

To summarize, the only case when we have slice uniformity is the "circle" $x_1^2 + x_2^2 = 0$ in $\mathbb{F}_p^2$, for $p \equiv 3 \bmod 4$.

Exercise 1.36. We note that 1, 2, and 4 are the quadratic residues modulo 7, and that $(-7/p) = (p/7)$ by the law of quadratic reciprocity (22). Thus, $p \equiv 1, 2, 4 \bmod 7$ is equivalent to $(-7/p) = 1$.

By Lemma 1.28, $(-7/p) = 1$ is necessary for representing $p = A^2 + 7B^2$ for some integers A and B. Our aim is to show that it is also sufficient, so assume $(-7/p) = 1$ in what follows. By Lemma 1.29, there exist integers A and B such that $A^2 + 7B^2 = kp$ for some $k \in \{1,\ldots,7\}$. In particular, $A^2 \equiv kp \bmod 7$. If $A \equiv 0 \bmod 7$, then a fortiori $k = 7$; upon writing $A = 7A'$, the relation $A^2 + 7B^2 = 7p$ becomes $7A'^2 + B^2 = p$ and we are done. If $A \not\equiv 0 \bmod 7$, then $A^2 \equiv kp \bmod 7$ implies that $(kp/7) = 1$. As $(p/7) = 1$, this means that $(k/7) = 1$, that is, $k \in \{1, 2, 4\}$. We will rule out the cases $k = 2$ and $k = 4$, leaving $k = 1$ as the only option. Thus, once again, $A^2 + 7B^2 = p$ for some integers A and B.

The case $k = 2$ is impossible since the relation $A^2 + 7B^2 = 2p$ fails modulo 4: as A^2 and B^2 are 0 or 1 mod 4, the left-hand side is 0, 1, or 3 mod 4; the right-hand side, however, is 2 mod 4. Similarly, the case $k = 4$ is impossible since the relation $A^2 + 7B^2 = 4p$ fails modulo 8: as A^2 and B^2 are 0 or 1 mod 8, the left-hand side is 0, 1, or 7 mod 8; but the right-hand side is 4 mod 8.

Chapter 2

Exercise 2.13. The equation

$$\frac{x}{y} + \frac{y}{z} + \frac{z}{x} = 3 \tag{148}$$

turns into

$$\frac{1}{y} + \frac{y}{z} + z = 3 \tag{$*$}$$

upon rescaling $y := yx$ and $z := zx$. Since $(*)$ is x-free, the solution count for (148) is $(p-1)N_*$, where N_* is the number of solutions $(y, z) \in \mathbb{F}_p^* \times \mathbb{F}_p^*$ to $(*)$. We rewrite $(*)$ as

$$z + y^2 + yz^2 = 3yz \tag{$**$}$$

which we interpret as a quadratic equation in z, for each $y \in \mathbb{F}_p^*$. The discriminant is

$$\Delta = (1 - 3y)^2 - 4y^3 = (y - 1)^2(1 - 4y),$$

so $(**)$ has $1 + \sigma(\Delta) = 1 + \sigma(1 - 4y)$ solutions whenever $y \neq 0, 1$, respectively one solution when $y = 1$. Note that a solution z to $(**)$ automatically satisfies $z \neq 0$ whenever $y \in \mathbb{F}_p^*$. Thus,

$$N_* = 1 + \sum_{y \neq 0,1} \big(1 + \sigma(1 - 4y)\big) = p - 1 + \sum_{y \neq 0,1} \sigma(1 - 4y)$$

$$= p - 1 - \sigma(1) - \sigma(-3) = p - 2 - \sigma(-3).$$

We conclude that the number of solutions to (148) is $(p-1)\big(p-2-\sigma(-3)\big)$.

Exercise 2.14. For each $x, y \in \mathbb{F}_p$, we view the equation

$$(x + y + z + 1)^2 = axyz \tag{149}$$

as a quadratic equation in z. An equation of the form $(\alpha + z)^2 = \beta z$ has discriminant $\Delta = \beta^2 - 4\alpha\beta$; consequently, it has $1 + \sigma(\Delta) = 1 + \sigma(\beta)\sigma(\beta - 4\alpha)$ solutions. So, the Equation (149) has $1 + \sigma(axy)\sigma(axy - 4x - 4y - 4)$ solutions for each $x, y \in \mathbb{F}_p$.

Overall, the number of solutions to the Equation (149) is

$$N = \sum_{x,y \in \mathbb{F}_p} \Big(1 + \sigma(axy)\sigma(axy - 4x - 4y - 4)\Big)$$

$$= p^2 + \sum_{x \in \mathbb{F}_p} \sigma(ax) \sum_{y \in \mathbb{F}_p} \sigma(y)\sigma((ax - 4)y - 4(x + 1)).$$

In order to compute the inner-most sum, we note the formula

$$\sum_{z \in \mathbb{F}_p} \sigma(y)\sigma(\alpha y - \beta) = \sigma(\alpha)\big(\llbracket \beta = 0 \rrbracket p - 1\big). \tag{150}$$

Indeed, if $\alpha = 0$ then (150) clearly holds, as both sides vanish. If $\alpha \neq 0$, then (150) follows from (35). We then have

$$N = p^2 + \sum_{x \in \mathbb{F}_p} \sigma(ax)\sigma(ax - 4)\big(\llbracket x = -1 \rrbracket p - 1\big)$$

$$= p^2 + \sum_{x \in \mathbb{F}_p} \sigma\big(x(x - 4a^{-1})\big)\big(\llbracket x = -1 \rrbracket p - 1\big)$$

$$= p^2 + \sigma(1 + 4a^{-1})p - \sum_{x \in \mathbb{F}_p} \sigma\big(x(x - 4a^{-1})\big)$$

$$= p^2 + \sigma(1 + 4a^{-1})p + 1.$$

In the last step, we have used (35) once again.

Exercise 2.15. Firstly, we count the solutions (x, y, z, t) with $t = 0$. The given equation becomes $x^2 y + y^2 z = 0$. There are p^2 solutions in which $y = 0$, respectively, $(p-1)p$ solutions in which $y \neq 0$. So, there are $2p^2 - p$ solutions (x, y, z, t) with $t = 0$.

Secondly, we count the solutions (x, y, z, t) with $t \neq 0$. After rescaling $x := tx$, $y := ty$, $z := tz$, the given equation turns into the t-free equation

$$x^2 y + y^2 z + z^2 + cx = 0. \tag{151}$$

Hence, the number of solutions (x, y, z, t) with $t \neq 0$ is $(p-1)N(c)$, where $N(c)$ denotes the number of solutions (x, y, z) to (151).

We view (151) as a quadratic equation in z; for each $x, y \in \mathbb{F}_p$, it has $1 + \sigma(y^4 - 4(x^2 y + cx))$ solutions. Thus,

$$N(c) = \sum_{x, y \in \mathbb{F}_p} \left(1 + \sigma\big(y^4 - 4(x^2 y + cx)\big) \right)$$

$$= p^2 + \sigma(-1) \sum_{y \in \mathbb{F}_p} \sum_{x \in \mathbb{F}_p} \sigma\left(yx^2 + cx - \frac{y^4}{4} \right).$$

The inner sum vanishes for $y = 0$. For $y \neq 0$, the inner sum equals $-\sigma(y)$ if $c^2 + y^5 \neq 0$, respectively, $\sigma(y)(p-1)$ if $c^2 + y^5 = 0$. Overall, the formula $\sigma(y)(p[\![y^5 = -c^2]\!] - 1)$ captures all cases. We continue

$$N(c) = p^2 + \sigma(-1) \sum_{y \in \mathbb{F}_p} \sigma(y)\big(p[\![y^5 = -c^2]\!] - 1\big)$$

$$= p^2 + p \sum_{y \in \mathbb{F}_p} \sigma(-y)[\![y^5 = -c^2]\!] = p^2 + p \sum_{y \in \mathbb{F}_p} \sigma(y)[\![y^5 = c^2]\!].$$

We observe that $\sigma(y) = 1$ whenever $y^5 = c^2$. Thus,

$$N(c) = p^2 + pn(c), \qquad n(c) := \#\{y \in \mathbb{F}_p : y^5 = c^2\}.$$

To conclude, the desired number of solutions (x, y, z, t) is

$$2p^2 - p + (p-1)N(c) = p^3 + (p^2 - p)(n(c) + 1).$$

As for the count $n(c)$, it can be made explicit as follows:

$$n(c) = \begin{cases} 1 & \text{if } p \not\equiv 1 \bmod 5, \\ 0 & \text{if } p \equiv 1 \bmod 5, c \notin (\mathbb{F}_p^*)^5, \\ 5 & \text{if } p \equiv 1 \bmod 5, c \in (\mathbb{F}_p^*)^5. \end{cases}$$

Exercise 2.16. We adapt the argument given for Exercise 1.34. The change of variables $x := x + b/4$, $y := y + b/4$, $z := z + b/4$, $t := t + b/4$ leads to the equivalent system

$$\begin{cases} (x^3 + y^3 + z^3 + t^3) + \dfrac{3b}{4}(x^2 + y^2 + z^2 + t^2) = a - \dfrac{b^3}{16}, \\[2mm] x + y + z + t = 0. \end{cases}$$

By using the second equation, we seek to eliminate the variable t from the first equation. We write

$$x^3 + y^3 + z^3 + t^3 = x^3 + y^3 + z^3 - (x + y + z)^3$$
$$= -3(x + y)(y + z)(z + x),$$

and

$$x^2 + y^2 + z^2 + t^2 = x^2 + y^2 + z^2 + (x + y + z)^2$$
$$= (x + y)^2 + (y + z)^2 + (z + x)^2.$$

Thus, by setting $Z := x + y$, $X := y + z$, $Y := z + x$, we reach the equation

$$-3XYZ + \frac{3b}{4}(X^2 + Y^2 + Z^2) = a - \frac{b^3}{16}.$$

When $b = 0$, this amounts to $-3XYZ = a$; there are $(p-1)^2$ solutions when $a \neq 0$, respectively $3p^2 - 3p + 1$ solutions when $a = 0$.

When $b \neq 0$, we make yet another change of variables–the rescaling $X := bX/4$, $Y := bY/4$, $Z := bZ/4$. We arrive at

$$X^2 + Y^2 + Z^2 - XYZ = \frac{64a - 4b^3}{3b^3}.$$

This is a generalized Markov cubic, as discussed in Example 2.7. Using the solution count (38), and performing some simplifications, we reach the conclusion that there are

$$p^2 + 1 + \big(3\sigma(3b) + \sigma(16a - b^3)\big)\sigma(4a - b^3)p$$

solutions.

Exercise 2.17. A simplifying assumption goes a long way in this exercise. We aim to show that there exist $A, B \in \mathrm{SL}_2(\mathbb{F}_p)$ such that $\operatorname{tr} A = a$,

$\operatorname{tr} B = b$, and

$$AB = \begin{pmatrix} c & 1 \\ -1 & 0 \end{pmatrix}.$$

The matrix we prescribed for the product AB is the simplest way to achieve determinant 1 and trace c. Now put

$$A = \begin{pmatrix} a - x & y \\ z & x \end{pmatrix},$$

so that $\operatorname{tr} A = a$ is achieved. We have

$$B = A^{-1} \begin{pmatrix} c & 1 \\ -1 & 0 \end{pmatrix} = \begin{pmatrix} x & -y \\ -z & a - x \end{pmatrix} \begin{pmatrix} c & 1 \\ -1 & 0 \end{pmatrix} = \begin{pmatrix} cx + y & * \\ * & -z \end{pmatrix}.$$

Thus, imposing $\operatorname{tr} B = b$ determines $z = cx + y - b$. The only requirement left to fulfill is that $\det A = 1$. That turns into the equation $(a - x)x - y(cx + y - b) = 1$, that is

$$x^2 + y^2 + cxy - ax - by + 1 = 0.$$

This is the most general quadratic equation in two variables, and it turns out to be solvable for most choices of coefficients a, b, c. A translation $x := x - (cy - a)/2$ turns it into

$$x^2 + \left(1 - \frac{c^2}{4}\right) y^2 - \left(b - \frac{ac}{2}\right) y + \left(1 - \frac{a^2}{4}\right) = 0.$$

If $c^2 \neq 4$, the above equation takes the form $x^2 + (1 - c^2/4)(y + \delta)^2 = \gamma$, which we know to be solvable. If $c^2 = 4$, the equation remains solvable except possibly when $b = ac/2$, and $a^2 \neq 4$.

The troublesome case, when $c^2 = 4$ and $a^2 \neq 4$, can be cleverly dealt with by noticing that a, b, c play a symmetric role. Indeed, assume that (A, B) is a solution for the tracial triple (a, b, c). Then (B, A) is a solution for the tracial triple (b, a, c) and, more interestingly, (AB, B^{-1}) is a solution for the tracial triple (c, b, a).

Exercise 2.18. Let $V \subseteq \mathbb{F}_p$ be the image of the cubic map $x \mapsto x^3 + ax + b$; according to (42), $\#V \geq (2p - 1)/3$. Let $S \subseteq \mathbb{F}_p$ be the image of the quadratic map $y \mapsto y^2$; we know that $\#S = (p + 1)/2$. We can then give a lower bound for the intersection of S with V:

$$\#(S \cap V) = \#S + \#V - \#(S \cup V) \geq \frac{p+1}{2} + \frac{2p-1}{3} - p = \frac{p+1}{6}.$$

Each point of intersection between S with V yields some solution to the given equation, $y^2 = x^3 + ax + b$; however, each non-zero point of intersection yields at least two solutions. So, the number of solutions is at least

$$2\big(\#(S \cap V) - 1\big) + 1 \geq \frac{p-2}{3}.$$

Exercise 2.19. We adapt the proof of Theorem 2.10. For $k = 1, 2, 3, 4$ we let m_k denote the number of $x \in \mathbb{F}_p$ with the property that $f(y) = f(x)$ has k distinct solutions as an equation in y. Then

$$\#f(\mathbb{F}_p) = m_1 + \frac{m_2}{2} + \frac{m_3}{3} + \frac{m_4}{4}.$$

We will determine m_1, m_3, and m_4. As for m_2, it is determined by the formula $p = m_1 + m_2 + m_3 + m_4$. Thus

$$\#f(\mathbb{F}_p) = \frac{p}{2} + \frac{m_1}{2} - \frac{m_3}{6} - \frac{m_4}{4}.$$

For each $x \in \mathbb{F}_p$, the equation $f(y) = f(x)$ amounts to the equation

$$(y^2 - x^2)(y^2 + x^2 + a) = 0. \tag{$\ddagger$}$$

When $x = 0$, ($\ddagger$) becomes $y^2(y^2 + a) = 0$; there are $2 + \sigma(-a)$ distinct solutions. When $x^2 = -a/2$, ($\ddagger$) becomes $y^2 = -a/2$; there are 2 distinct solutions, $y = \pm x$. When $x^2 \neq 0, -a/2$, ($\ddagger$) has $3 + \sigma(-x^2 - a)$ distinct solutions; thus, either 2 or 4 solutions, except when $x^2 = -a$, in which case we get 3 solutions.

We deduce that

$$m_1 = \begin{cases} 1 & \text{if } \sigma(-a) = -1 \\ 0 & \text{if } \sigma(-a) = 1 \end{cases} = \frac{1 - \sigma(-a)}{2}$$

and

$$m_3 = \begin{cases} 0 & \text{if } \sigma(-a) = -1 \\ 3 & \text{if } \sigma(-a) = 1 \end{cases} = \frac{3(1 + \sigma(-a))}{2}$$

and

$$m_4 = \#\{x : x^2 \neq 0, -a/2, \ \sigma(-x^2 - a) = 1\}.$$

The latter count we evaluate as follows:

$$m_4 = \sum_{x\,:\,x^2 \neq 0,\,-a/2,\,-a} \frac{1 + \sigma(-x^2 - a)}{2}$$

$$= \sum_{x \in \mathbb{F}_p} \frac{1 + \sigma(-x^2 - a)}{2} - \sum_{c=0,\,-a/2,\,-a} \left(\sum_{x\,:\,x^2 = c} \frac{1 + \sigma(-x^2 - a)}{2} \right).$$

The left-hand sum evaluates as $(p-1)/2$. The inner-most sum on the right evaluates as $(1 + \sigma(c))(1 + \sigma(-c - a))/2$. We are led to

$$m_4 = \frac{p - \sigma(-1)}{2} - (1 + \sigma(-a)) - (1 + \sigma(-a/2)).$$

Now, it is a simple matter to calculate that

$$\#f(\mathbb{F}_p) = \frac{3p + 4 + \sigma(-1)}{8} + \frac{\sigma(2) - 1}{4} \cdot \sigma(-a).$$

Below we tabulate explicit formulas for $\#f(\mathbb{F}_p)$; explicit forms for $\lceil 3p/8 \rceil$; finally, the link between the two, which completes the argument.

$p \bmod 8$	1	3	5	7
$\#f(\mathbb{F}_p)$	$\dfrac{3p+5}{8}$	$\dfrac{3p+3}{8} + \dfrac{\sigma(a)}{2}$	$\dfrac{3p+5}{8} - \dfrac{\sigma(a)}{2}$	$\dfrac{3p+3}{8}$
$\left\lceil \dfrac{3p}{8} \right\rceil$	$\dfrac{3p+5}{8}$	$\dfrac{3p+7}{8}$	$\dfrac{3p+1}{8}$	$\dfrac{3p+3}{8}$
$\#f(\mathbb{F}_p)$	$\left\lceil \dfrac{3p}{8} \right\rceil$	$\left\lceil \dfrac{3p}{8} \right\rceil - \dfrac{1 - \sigma(a)}{2}$	$\left\lceil \dfrac{3p}{8} \right\rceil + \dfrac{1 - \sigma(a)}{2}$	$\left\lceil \dfrac{3p}{8} \right\rceil$

Chapter 3

Exercise 3.26. Fix $b \in \mathbb{F}_p^*$. As already pointed out in Example 2.11, the irreducibility of the polynomial $X^3 + aX + b$ amounts to $x^3 + ax + b = 0$ having no roots. This, in turn, means that $-a$ cannot be in the image of the rational map $f : \mathbb{F}_p^* \to \mathbb{F}_p$ given by

$$f(x) = \frac{x^3 + b}{x}.$$

The heart of the matter, then, is to count the number of values of the map f. We claim that

$$\#f(\mathbb{F}_p^*) = \frac{2p - 2 - \varphi_3(4b)}{3}. \tag{$\dagger$}$$

Thus, the number of choices for $a \in \mathbb{F}_p$ which render the polynomial $X^3 + aX + b$ irreducible over $\mathbb{F}_p$ is

$$p - \#f(\mathbb{F}_p^*) = \frac{2p + 2 + \varphi_3(4b)}{3}.$$

Turning to the proof of ($\dagger$), we follow the blueprint of Theorem 2.10 once again. The map f is at most three-to-one. For $k = 1, 2, 3$, let m_k denote the number of $x \in \mathbb{F}_p^*$ with the property that $f(y) = f(x)$ has k distinct solutions as an equation in y. We have $m_1 + m_2 + m_3 = p - 1$, and

$$\#f(\mathbb{F}_p^*) = m_1 + \frac{m_2}{2} + \frac{m_3}{3} = \frac{p - 1}{3} + \frac{2m_1}{3} + \frac{m_2}{6}.$$

Rewriting the equation $f(y) = f(x)$, we find that we have to count the number of solutions y to the quadratic equation

$$xy^2 + x^2 y - b = 0 \tag{$*$}$$

which are different from x. We note that x is a solution to ($*$) when $x^3 = b/2$; let n denote the number of solutions to the latter equation.

The discriminant of ($*$) is $\Delta(x) = x^4 + 4bx$. We note that the equation $\Delta(x) = 0$ has n solutions in $\mathbb{F}_p^*$; indeed, $x^3 = -4b$ has n solutions. We also note that, when $\Delta(x) = 0$, x cannot be a solution to ($*$); for $\Delta(x) = 0$ and $x^3 = b/2$ are incompatible.

Now, m_1 counts those $x \in \mathbb{F}_p^*$ for which ($*$) has no solution, so

$$m_1 = \sum_{x \in \mathbb{F}_p^*} [\![\sigma(\Delta(x)) = -1]\!] = \sum_{x \in \mathbb{F}_p^*} \left(\frac{1 - \sigma(\Delta(x))}{2} - \frac{[\![\Delta(x) = 0]\!]}{2} \right)$$

$$= \frac{p - 1}{2} - \frac{1}{2} \sum_{x \in \mathbb{F}_p^*} \sigma(x^4 + 4bx) - \frac{n}{2} = \frac{p - 1 - \varphi_3(4b) - n}{2}.$$

The multiplicity m_2 counts those $x \in \mathbb{F}_p^*$ for which either ($*$) has a unique solution, different from x, or ($*$) has two solutions, one of which is x. The first case occurs when $\Delta(x) = 0$; the unique solution is different from x. This case contributes n towards m_2. The second case occurs when

$\sigma(\Delta(x)) = 1$ and $x^3 = b/2$. But the latter already makes $\Delta(x) = 9x^4$, a square indeed. This case also contributes n towards m_2, so

$$m_2 = 2n.$$

We conclude that

$$\#f(\mathbb{F}_p^*) = \frac{p-1}{3} + \frac{2m_1}{3} + \frac{m_2}{6} = \frac{p-1}{3} + \frac{p-1-\varphi_3(4b)-n}{3} + \frac{n}{3}$$
$$= \frac{2p-2-\varphi_3(4b)}{3},$$

so (†) holds. As a sanity check, let us certify that $2p-2-\varphi_3(4b) \equiv 0 \bmod 3$. For if $p \equiv 1 \bmod 3$, then $\varphi_3(4b) \equiv 0 \bmod 3$ by part (ii) of Lemma 3.3; if $p \equiv 2 \bmod 3$, then $\varphi_3(4b) = \varphi_1(4b) = -1$ by (55).

Exercise 3.27. When $n = (p-1)/2$, we have

$$\varphi_n(a) = \sum_{x \in \mathbb{F}_p^*} \sigma(x)\sigma\big(x^{(p-1)/2} + a\big).$$

When $x \in \mathbb{F}_p^*$ is a square, $\sigma(x)\sigma(x^{(p-1)/2} + a) = \sigma(1+a)$; when $x \in \mathbb{F}_p^*$ is not a square, $\sigma(x)\sigma(x^{(p-1)/2} + a) = -\sigma(-1+a)$. Overall

$$\varphi_n(a) = \frac{p-1}{2}\big(\sigma(a+1) - \sigma(a-1)\big).$$

The possible values of $\varphi_n(a)$ are therefore $\pm(p-1)$ or 0 for all $a \in \mathbb{F}_p$, except for $a = \pm 1$ in which case $\varphi_n(a) = \pm(p-1)/2$.

Exercise 3.28. By definition, $C = \psi_3(1)$ and $D = \sigma(a)\psi_3(a)$. We turn from ψ_3-sums to φ_3-sums: $C = 1+\varphi_3(1)$ and $D = 1+\varphi_3(a^{-1})$. Thus $C = 1+w_0$, and $D = 1+w_1$ or $D = 1+w_2$; yet even more convenient is to recast $C = v_0$, and $D = v_1$ or $D = v_2$, in the notation used in the proof of Theorem 3.16. Therein, we have seen that $v_0 + v_1 + v_2 = 0$ and $v_0^2 + v_1^2 + v_2^2 = 6p$. Expressing v_2 as $v_2 = -(v_0 + v_1)$ leads to $v_0^2 + v_1^2 + v_0v_1 = 3p$; likewise, $v_0^2 + v_2^2 + v_0v_2 = 3p$. We conclude that $C^2 + D^2 + CD = 3p$.

Exercise 3.29. Let $a \in \mathbb{F}_p^*$. The easier bit is showing that $|1+\psi_4(a)| \le 2\sqrt{p}$. Indeed, $1 + \psi_4(a) = \varphi_2(a)$, and $|\varphi_2(a)| \le 2\sqrt{p}$ by (69).

Proving that $|1 + \varphi_3(a)| \le 2\sqrt{p}$ requires more detail. A similar attempt would be to use $1 + \varphi_3(a) = \sigma(a)\psi_3(a^{-1})$, which gives $|1 + \varphi_3(a)| = |\psi_3(a^{-1})|$; unfortunately, we only know that $|\psi_3(a^{-1})| \le 3\sqrt{p}$ from (69). We argue instead as follows. If $p \equiv 2 \bmod 3$ then $\varphi_3(a) = \varphi_1(a) = -1$, and the desired bound obviously holds. If $p \equiv 1 \bmod 3$ then, from part (ii) of Theorem 3.17, we know that

$$|1 + \varphi_3(a)| \le \max\left\{2|A|, |A| + 3|B|\right\},$$

where $p = A^2 + 3B^2$. Clearly, $|A| \le \sqrt{p}$, so $2|A| \le 2\sqrt{p}$. By the Cauchy–Schwarz inequality, we have $(|A|+3|B|)^2 \le (1+3)(A^2+3B^2) = 4p$; therefore $|A| + 3|B| \le 2\sqrt{p}$ as well.

Exercise 3.30. We need to analyze the modular condition $\psi_3(a) \equiv -1$ $\bmod\ p$. When $p \equiv 2 \bmod 3$, we have $\psi_3(a) = \psi_1(a) = 0$ for all $a \in \mathbb{F}_p^*$. Therefore, $p \equiv 1 \bmod 3$.

As $\psi_3(a) \in [-p, p]$, we either have $\psi_3(a) = -1$, or $\psi_3(a) = p - 1$. Note, however, that the latter case is impossible, since $\psi_3(a) \equiv \sigma(a) = \pm 1$ $\bmod\ 3$ by part (i) of Lemma 3.3. Therefore, $\psi_3(a) = -1$. We also infer that $\sigma(a) = -1$, that is to say, a is a non-square.

The formula $\psi_3(a) = \sigma(a)(1 + \varphi_3(a^{-1}))$ turns $\psi_3(a) = -1$ into $\varphi_3(a^{-1}) = 0$. Part (ii) of Theorem 3.17 offers two possibilities for $\varphi_3(a^{-1})$: either $2A - 1$ or $-A - 1 \pm 3B$, where $p = A^2 + 3B^2$ and $A \equiv -1 \bmod 3$. Clearly, we must be in the latter case. Thus, $A + 1 = \pm 3B$, and a^{-1} is a non-cube; equivalently, a is a non-cube. Set $A = 3m - 1$ for some positive integer m; then $B = \pm m$. It follows that

$$p = A^2 + 3B^2 = (3m - 1)^2 + 3m^2 = 12m^2 - 6m + 1.$$

We have thus reached the necessary conditions that p is a prime of the form $12m^2 - 6m + 1$, and a is neither a square nor a cube in $\mathbb{F}_p^*$.

Unfortunately, we cannot assert that these two conditions are sufficient. From $p = 12m^2 - 6m + 1$ we do find that $A = 3m - 1$ and $B = \pm m$; but from $\varphi_3(a^{-1}) = -A - 1 \pm 3B$ we only know that $\varphi_3(a^{-1}) = 0$ or $\varphi_3(a^{-1}) = 6m$.

Exercise 3.31. (i) We adapt the proof of the second moment formula (68). We have

$$\sum_{a \in \mathbb{F}_p} \varphi_n(a)\varphi_n(ca) = \sum_{a \in \mathbb{F}_p} \left(\sum_{x \in \mathbb{F}_p} \sigma(x^{n+1} + ax)\right)\left(\sum_{y \in \mathbb{F}_p} \sigma(y^{n+1} + cay)\right)$$

$$= \sum_{x,y \in \mathbb{F}_p} \sigma(cxy) \sum_{a \in \mathbb{F}_p} \sigma(a + x^n)\sigma(a + y^n c^{-1}).$$

The inner sum equals -1, except when $x^n = y^n c^{-1}$ in which case it equals $p - 1$. The latter case can only occur for $x = y = 0$ since c is not an nth power. So, in fact,

$$\sum_{a \in \mathbb{F}_p} \varphi_n(a)\varphi_n(ca) = - \sum_{x,y \in \mathbb{F}_p} \sigma(cxy) = 0.$$

(ii) For each integer s, the twisted periodicity (75) implies that the sequence $\{w_k w_{k+s}\}_k$ is n-periodic regardless of the parity of n. Therefore,

$$\sum_{a \in \mathbb{F}_p^*} \varphi_n(a)\varphi_n(g^s a) = \sum_{k=0}^{p-2} \varphi_n(g^k)\varphi_n(g^{k+s})$$

$$= \sum_{k=0}^{p-2} w_k w_{k+s} = \frac{p-1}{n} \sum_{k=0}^{n-1} w_k w_{k+s}.$$

Now, if s is in the range $1, \dots, n-1$, then g^s is not an nth power; by part (i), the left-hand side equals $-\varphi_n(0)^2$. As $\varphi_n(0) = 0$ if n is even, respectively, $\varphi_n(0) = p - 1$ if n is odd, we deduce that

$$\sum_{k=0}^{n-1} w_k w_{k+s} = \begin{cases} -n(p-1) & \text{if } n \text{ is odd} \\ 0 & \text{if } n \text{ is even} \end{cases}.$$

Exercise 3.32. We borrow an argument from the proof Lemma 3.19. Namely, we evaluate

$$N_\epsilon = \#\{x \in \mathbb{F}_p : \sigma(x-1) = \sigma(x) = \sigma(x+1) = \epsilon\},$$

where $\epsilon = \pm 1$, by means of the quadratic character sum

$$S_\epsilon = \sum_{x \in \mathbb{F}_p} \big(\sigma(x-1) + \epsilon\big)\big(\sigma(x) + \epsilon\big)\big(\sigma(x+1) + \epsilon\big).$$

On the one hand

$$S_\epsilon = 8\epsilon N_\epsilon + 3\epsilon + 2 + (\epsilon + 2)\sigma(-1) + (2\epsilon + 1)\sigma(2) + \sigma(-2).$$

Here, the main term, $8\epsilon N_\epsilon$, cumulates the contribution to S_ϵ of those $x \in \mathbb{F}_p$ for which $\sigma(x-1) = \sigma(x) = \sigma(x+1) = \epsilon$; the remaining crumbs come from $x = -1, 0, 1$; terms corresponding to all other $x \in \mathbb{F}_p$ vanish.

On the other hand

$$S_\epsilon = \epsilon(p-3) + \varphi_2(-1)$$

by breaking S_ϵ into several quadratic character sums. The Jacobsthal sum $\varphi_2(-1)$ accounts for $\sum_{x \in \mathbb{F}_p} \sigma(x-1)\sigma(x)\sigma(x+1)$.

We deduce a formula for N_ϵ by equating the two evaluations of S_ϵ. When $p \equiv 3 \bmod 4$, we have $\sigma(-1) = -1$ and $\varphi_2(-1) = 0$, we obtain

$$8N_\epsilon = p - 5 - 2\sigma(2).$$

In this case, N_ϵ is independent of ϵ.

When $p \equiv 1 \bmod 4$, we obtain

$$8N_\epsilon = (p - 3) + \epsilon\varphi_2(-1) - 2(\epsilon + 1)(2 + \sigma(2)).$$

In this case, N_ϵ is independent of ϵ precisely when $\varphi_2(-1) = 2(2 + \sigma(2))$. As $\varphi_2(-1) = (-1)^{(p-1)/4}\varphi_2(1) = (-1)^{(p-1)/4} \cdot 2A_2(p)$, this is equivalent to

$$A_2(p) = (-1)^{(p-1)/4}(2 + \sigma(2)) = \begin{cases} 3 & \text{if } p \equiv 1 \bmod 8, \\ -1 & \text{if } p \equiv 5 \bmod 8. \end{cases}$$

Now, $A_2(p) = 3$ and $p \equiv 1 \bmod 8$ precisely when $p = B^2 + 9$ with $B \equiv 0 \bmod 4$; in other words, $p = 16m^2 + 9$ for some integer m. Similarly, $A_2(p) = -1$ and $p \equiv 5 \bmod 8$ precisely when $p = B^2 + 1$ with $B \equiv 2 \bmod 4$; that is, $p = 16m^2 + 16m + 5$ for some integer m.

Exercise 3.33. We first express the double variable sum as a Jacobsthal sum. Isolating the part corresponding to $y = 0$, and making a change of variable $x := xy$ in the part where $y \neq 0$, we get:

$$\sum_{x,y \in \mathbb{F}_p} \sigma(x^6 - y^6) = \sum_{x \in \mathbb{F}_p} \sigma(x^6) + \sum_{y \in \mathbb{F}_p^*} \sum_{x \in \mathbb{F}_p} \sigma(x^6 - y^6)$$

$$= (p - 1) + \sum_{y \in \mathbb{F}_p^*} \sum_{x \in \mathbb{F}_p} \sigma(x^6 - 1)$$

$$= (p - 1) + (p - 1)\psi_6(-1).$$

Next, we evaluate $\psi_6(-1)$ by stringing together a few facts about Jacobsthal sums. If $p \not\equiv 1 \bmod 3$, then by (55) we have $\psi_6(-1) = \psi_2(-1) = -1$. Assume $p \equiv 1 \bmod 3$ in what follows. Then

$$\psi_6(-1) = \psi_3(-1) + \varphi_3(-1) = \sigma(-1) + (\sigma(-1) + 1)\varphi_3(-1),$$

the first equality owing to (54), and the second to (60). If $p \not\equiv 1 \bmod 4$, then $\sigma(-1) = -1$ and so $\psi_6(-1) = -1$ once again. If $p \equiv 1 \bmod 4$, then

$\psi_6(-1) = 1 + 2\varphi_3(-1)$. As $\varphi_3(-1) = \varphi_3(1)$ by the sign-change rule (63), and $\varphi_3(1) = 2A_3(p) - 1$, we conclude that

$$\psi_6(-1) = \begin{cases} 4A_3(p) - 1 & \text{if } p \equiv 1 \bmod 12, \\ -1 & \text{otherwise.} \end{cases}$$

All in all

$$\sum_{x,y\in\mathbb{F}_p} \sigma(x^6 - y^6) = \begin{cases} 4(p-1)A_3(p) & \text{if } p \equiv 1 \bmod 12, \\ 0 & \text{otherwise.} \end{cases}$$

Exercise 3.34. Throughout, congruences are modulo p.

(i) Wilson's congruence gives

$$-1 \equiv (p-1)! \equiv (-1)^k (p-1-k)! \, k!$$

There is, of course, a similar modular identity for j. Thus,

$$\binom{p-1-j}{p-1-k} = \frac{(p-1-j)!}{(k-j)! \, (p-1-k)!} \equiv \frac{(-1)^k \, k!}{(k-j)! \, (-1)^j \, j!} = (-1)^{k-j} \binom{k}{j}.$$

(ii) We note that

$$\binom{\frac{7}{8}(p-1)}{\frac{1}{2}(p-1)} = \binom{(p-1) - \frac{1}{8}(p-1)}{(p-1) - \frac{1}{2}(p-1)} \equiv (-1)^{\frac{1}{2}(p-1) - \frac{1}{8}(p-1)} \binom{\frac{1}{2}(p-1)}{\frac{1}{8}(p-1)}.$$

Here, $\frac{1}{2}(p-1)$ is even. By part (iii) of Theorem 3.22, we conclude that

$$\binom{\frac{7}{8}(p-1)}{\frac{1}{2}(p-1)} \equiv -2A_4(p).$$

Exercise 3.35. Throughout, congruences are modulo p.

(i) We write

$$(-4)^k \frac{\left(\frac{p-1}{2}\right)!}{\left(\frac{p-1}{2} - k\right)!} = (-4)^k \left(\frac{p-1}{2} - k + 1\right)\left(\frac{p-1}{2} - k + 2\right) \cdots \left(\frac{p-1}{2}\right)$$

$$= (-2)^k (p - (2k-1))(p - (2k-3)) \cdots (p-1)$$

$$\equiv 2^k (2k-1)(2k-3) \cdots 1 = \frac{(2k)!}{k!}.$$

Dividing through by $k!$, which is invertible mod p as $k < p$, we get the desired congruence:

$$(-4)^k \binom{\frac{1}{2}(p-1)}{k} \equiv \binom{2k}{k}.$$

(ii) Taking $k = \frac{1}{8}(p-1)$ in the above congruence, we get

$$\binom{\frac{1}{4}(p-1)}{\frac{1}{8}(p-1)} \equiv (-4)^{(p-1)/8} \binom{\frac{1}{2}(p-1)}{\frac{1}{8}(p-1)} \equiv -4^{(p-1)/8} \cdot 2A_4(p).$$

The last congruence comes from part (iii) of Theorem 3.22. Next, we write $4^{(p-1)/8} = 2^{(p-1)/4} = (-1)^{(A_4(p)+1)/4}$, as seen in Example 3.21. All in all,

$$\binom{\frac{1}{4}(p-1)}{\frac{1}{8}(p-1)} \equiv (-1)^{(A_4(p)+5)/4} \cdot 2A_4(p).$$

Chapter 4

Exercise 4.13. The given equation can be rewritten as

$$yx^2 - (2y^2 - c)x - cy = 0.$$

If $y = 0$, then necessarily $x = 0$. For each $y \neq 0$, the above quadratic equation in x has $1 + \sigma(4y^4 + c^2)$ solutions. So, the total number of solutions is

$$N = 1 + \sum_{y \in \mathbb{F}_p^*} \left(1 + \sigma(4y^4 + c^2)\right) = p - 1 + \sum_{y \in \mathbb{F}_p} \sigma(4y^4 + c^2).$$

(The given equation is asymmetric in x and y. Rewriting it as a quadratic equation in y, namely $2xy^2 - (x^2 - c)y - cx = 0$, leads to a more complicated quadratic character sum.) The latter sum is the Jacobsthal sum $\psi_4(c^2/4)$, and

$$\psi_4(c^2/4) = -1 + \varphi_2(c^2/4) = -1 + \sigma(c/2)\varphi_2(1).$$

We conclude that

$$N = p - 2 + \sigma(2c)\varphi_2(1).$$

Exercise 4.14. Let N_0, respectively N_{-1}, denote the number of solutions to the equations $x^4 + y^4 + z^4 = 0$ and $x^4 + y^4 + z^4 = -1$. Thanks to Example 4.10, we know that

$$N_0 = p^2 + 3(p-1)\varphi_2(1).$$

The number of solutions to $x^4 + y^4 + z^4 + t^4 = 0$ is, on the one hand, $N_0 + (p-1)N_{-1}$; on the other hand, according to Theorem 4.11, it is $p^3 + (5 + 12\epsilon)(p-1)p + \varphi_2(1)^2(p-1)$. Using the value of N_0 computed above, we deduce that

$$N_{-1} = p^2 + (5 + 12\epsilon)p + \varphi_2(1)^2 - 3\varphi_2(1).$$

Exercise 4.15. We take the convolutional viewpoint on computing the number N of solutions to $x^3 + y^3 + c(z^3 + t^3) = 0$:

$$N = \sum_{a\in\mathbb{F}_p} N(x^3 + y^3 = a)\, N(z^3 + t^3 = -ca).$$

By appealing to Theorem 4.5, we obtain

$$N = (3p - 2)^2 + \sum_{a\in\mathbb{F}_p^*} \left(p - 1 + \varphi_3(a/2)\right)\left(p - 1 + \varphi_3(-ca/2)\right)$$

$$= (3p - 2)^2 - 4(p-1)^2 + \sum_{a\in\mathbb{F}_p} \left(p - 1 + \varphi_3(a)\right)\left(p - 1 + \varphi_3(-ca)\right)$$

after the change of variable $a := 2a$, and using $\varphi_3(0) = p - 1$. Recalling also that $\sum_{a\in\mathbb{F}_p} \varphi_3(a) = 0$, by (66), we obtain after a short calculation that

$$N = p^3 + 3(p^2 - p) + \sum_{a\in\mathbb{F}_p} \varphi_3(a)\varphi_3(-ca).$$

Now,

$$\sum_{a\in\mathbb{F}_p} \varphi_3(a)\varphi_3(-ca) = \begin{cases} 3(p^2 - p) & \text{if } c \text{ is a cube in } \mathbb{F}_p^*, \\ 0 & \text{if } c \text{ is not a cube in } \mathbb{F}_p^*. \end{cases}$$

Indeed, the cube case follows from (68) and the periodicity (62); the non-cube case follows from part (i) of Exercise 3.31. In conclusion

$$N = \begin{cases} p^3 + 6(p^2 - p) & \text{if } c \text{ is a cube in } \mathbb{F}_p^*, \\ p^3 + 3(p^2 - p) & \text{if } c \text{ is not a cube in } \mathbb{F}_p^*. \end{cases}$$

Next, let N_1 denote the number of solutions to the equation $x^3 + c(y^3 + z^3) = 1$; analogously, we may define N_0 and N_{-1}. The number of solutions to the equation $x^3 + c(y^3 + z^3) + t^3 = 0$ satisfies the relation

$$N = N_0 + (p-1)N_{-1}.$$

In fact $N_{-1} = N_1$, by the correspondence $(x, y, z) \leftrightarrow (-x, -y, -z)$. On the other hand, for N_0 we know that

$$N_0 = p^2 + p - 1 + (p-1)\varphi_3(1/(2c))$$

by Example 4.7. Whence

$$N_1 = \frac{N - N_0}{p-1} = \begin{cases} p^2 + 6p - 1 - \varphi_3(1/2) & \text{if } c \text{ is a cube in } \mathbb{F}_p^*, \\ p^2 + 3p - 1 - \varphi_3(1/(2c)) & \text{if } c \text{ is not a cube in } \mathbb{F}_p^*. \end{cases}$$

Here, we have used the observation that $\varphi_3(1/(2c)) = \varphi_3(1/2)$ whenever c is a cube in $\mathbb{F}_p^*$.

Exercise 4.16. For $p \not\equiv 1 \bmod 4$, the equation $x^4 + y^4 = c$ has the same number of solutions as $x^2 + y^2 = c$, which is solvable for any p. Assume $p \equiv 1 \bmod 4$ in what follows. We may also assume that $c \neq 0$, since $x^4 + y^4 = 0$ is obviously solvable.

According to Theorem 4.8, the equation $x^4 + y^4 = c$ has

$$N(c) = p - 1 - 2\epsilon + 2\epsilon\varphi_2(c) + \sigma(c)\varphi_2(1)$$

solutions, where $\epsilon = \pm 1$. Then

$$N(c) \geq p - 3 - 2|\varphi_2(c)| - |\varphi_2(1)| \geq p - 3 - 6\sqrt{p}$$

as, thanks to the bound (69), we have $|\varphi_2(c)| \leq 2\sqrt{p}$ for each $c \in \mathbb{F}_p^*$. Now $p - 3 - 6\sqrt{p} > 0$, whenever $p > 41$. Thus, $N(c) > 0$, meaning that $x^4 + y^4 = c$ is solvable, whenever $p > 41$.

It remains to settle the range $29 < p \leq 41$. The assumption $p \equiv 1 \bmod 4$ isolates just two values in this range, $p = 37$ and $p = 41$. Consider the sum-of-squares representation $p = A^2 + B^2$, with the convention that A and B are positive, A is odd, and B is even. Part (i) of Theorem 3.17

says that $\varphi_2(1) = \pm 2A$ and $\varphi_2(c) \in \{\pm 2A, \pm 2B\}$. Now,

$$N(c) \geq p - 3 - 2|\varphi_2(c)| - |\varphi_2(1)| \geq p - 3 - 4\max\{A, B\} - 2A.$$

For $p = 41$, we have $A = 5$ and $B = 4$; for $p = 37$, we have $A = 1$ and $B = 6$. In both cases, we see that $N(c) > 0$, as desired.

We now let $p = 29$, and we determine the values of c for which $N(c) = 0$. Here, $\epsilon = -1$ and $\varphi_2(1) = 2A_2(p) = -10$, so

$$N(c) = 30 - 2\varphi_2(c) - 10\sigma(c).$$

Furthermore, $\varphi_2(c) \in \{\pm 10, \pm 4\}$. For $N(c) = 0$ to hold, c must be a square—say $c = d^2$. Then $\varphi_2(c) = \sigma(d)\varphi_2(1) = -10\sigma(d)$, so $N(c) = 20 + 20\sigma(d)$. We see that $N(c) = 0$ precisely when c is a square though not a fourth power; there are seven such elements in $\mathbb{F}_{29}$, namely, -1, 4, 5, 6, -7, 9, 13.

Exercise 4.17. If $p \not\equiv 1 \bmod 4$, then $x^4 + y^4 + az^2 = c$ has as many solutions as $x^2 + y^2 + az^2 = c$. By Theorem 1.19, that number is

$$N(c) = p^2 + \sigma(-ac)p = p^2 - \sigma(ac)p.$$

Assume $p \equiv 1 \bmod 4$ in what follows. By Theorem 4.8, the number of solutions to the equation $x^4 + y^4 = C$ is

$$p - 1 - 2\epsilon + 2\epsilon\varphi_2(C) + \sigma(C)\varphi_2(1) + (1 + 2\epsilon)p \cdot [\![C = 0]\!],$$

where $\epsilon = (-1)^{(p-1)/4}$. The point of the above formula is that it combines both cases, $C \neq 0$ and $C = 0$, into one functional expression. We use the above formula in order to count, for each $z \in \mathbb{F}_p$, the number of solutions to the equation $x^4 + y^4 = c - az^2$. Summing over $z \in \mathbb{F}_p$, we deduce that the number of solutions to $x^4 + y^4 + az^2 = c$ is

$$N(c) = p(p - 1 - 2\epsilon) + 2\epsilon S_1 + \varphi_2(1)S_2 + (1 + 2\epsilon)pS_3$$

where

$$S_1 = \sum_{z \in \mathbb{F}_p} \varphi_2(c - az^2), \quad S_2 = \sum_{z \in \mathbb{F}_p} \sigma(c - az^2), \quad S_3 = \sum_{z \in \mathbb{F}_p} [\![c - az^2 = 0]\!].$$

Clearly, $S_3 = 1 + \sigma(ac)$. Also $S_2 = \sigma(-a)(p[\![c = 0]\!] - 1)$; moreover, we note that $\sigma(-a) = \sigma(a)$. As for S_1, we evaluate it as follows:

$$\begin{aligned}
S_1 &= \sum_{z \in \mathbb{F}_p} \varphi_2(c - az^2) = \sum_{z \in \mathbb{F}_p} \sum_{u \in \mathbb{F}_p} \sigma(u)\sigma(u^2 + c - az^2) \\
&= \sum_{u \in \mathbb{F}_p} \sigma(u) \sum_{z \in \mathbb{F}_p} \sigma(u^2 + c - az^2) \\
&= \sum_{u \in \mathbb{F}_p} \sigma(u)\sigma(-a)\big(p[\![u^2 + c = 0]\!] - 1\big) \\
&= \sigma(a)p \sum_{u \in \mathbb{F}_p} \sigma(u)[\![u^2 = -c]\!].
\end{aligned}$$

The latter sum equals 0 when $c = 0$ or $c \notin (\mathbb{F}_p^*)^2$. When $c \in (\mathbb{F}_p^*)^2$, say $c = t^2$, the equation $u^2 = -c$ has two solutions: $u = \pm jt$, where $j^2 = -1$. So the latter sum equals $\sigma(jt) + \sigma(-jt) = 2\epsilon\sigma(t)$.

Combining all these calculations, we can deduce the following:

$$N(c) = \begin{cases}
p^2 + \sigma(a)(p - 1)\varphi_2(1) & \text{if } c = 0, \\
p^2 - \sigma(a)\big((1 + 2\epsilon)p + \varphi_2(1)\big) & \text{if } c \notin (\mathbb{F}_p^*)^2, \\
p^2 + \sigma(a)\big((1 + 2\epsilon + 4\sigma(t))p - \varphi_2(1)\big) & \text{if } c = t^2, t \in \mathbb{F}_p^*.
\end{cases}$$

Exercise 4.18. We aim to show that the system

$$\begin{cases}
x + y + z = 0 \\
xyz = a
\end{cases} \tag{152}$$

is solvable for every $a \in \mathbb{F}_p$, except for a few exceptions that we will determine. We may assume that $a \neq 0$, as (152) is clearly solvable for $a = 0$. Eliminating the variable z, the system amounts to the equation

$$x^2 y + xy^2 + a = 0.$$

There are no solutions for $x = 0$. For each $x \neq 0$, the above quadratic equation in y has $1 + \sigma(x^4 - 4ax)$ solutions. So, the number of solutions to the system (152) is

$$N(a) = \sum_{x \in \mathbb{F}_p^*} \big(1 + \sigma(x^4 - 4ax)\big) = p - 1 + \varphi_3(-4a).$$

We wish to argue that $N(a) > 0$. When $p \not\equiv 1 \bmod 3$, we have $\varphi_3(-4a) = \varphi_1(-4a) = -1$; whence $N(a) = p - 2 > 0$. Assume $p \equiv 1 \bmod 3$ in what follows. The bound (69) yields $|\varphi_3(-4a)| \le 3\sqrt{p}$. Consequently, $N(a) \ge p - 1 - 3\sqrt{p} > 0$ for all $p \ge 11$. An exception might arise when $p < 11$; necessarily $p = 7$ in view of our assumption that $p \equiv 1 \bmod 3$. In summary, (152) is solvable over all fields $\mathbb{F}_p$, except possibly when $p = 7$. (The same point is eventually reached by using a better bound, $|1 + \varphi_3(-4a)| \le 2\sqrt{p}$, afforded by Exercise 3.29.)

We determine when $N(a) = 0$ holds over $\mathbb{F}_7$. This means that $\varphi_3(-4a) = -(p-1)$, which in turn means that $x^4 - 4ax = x(x^3 - 4a)$ is a non-square whenever $x \ne 0$. When $x \ne 0$ is a square, we have $x^3 = 1$ and so we need $1 - 4a$ to be a non-square. When $x \ne 0$ is a non-square, we have $x^3 = -1$ and now we need $-1 - 4a$ to be a square. Thus, in $\mathbb{F}_7$, $1 - 4a \in \{3, 5, 6\}$, while $-1 - 4a \in \{1, 2, 4\}$. This determines $a = 3$ or $a = 4$. These two elements in $\mathbb{F}_7$ are the only two exceptions to the desired factorization.

Exercise 4.19. For each $c \in \mathbb{F}_p$, we let $N(c)$ denote the number of solutions in $\mathbb{F}_p^*$ to the equation

$$\frac{1}{x} + \frac{1}{y} + \frac{1}{z} = c(x + y + z). \tag{153}$$

By means of the scaling $(x, y, z) \mapsto (sx, sy, sz)$, we see that $N(s^2 c) = N(c)$ for any $s \in \mathbb{F}_p^*$. Consequently, there are only three values of interest: $N_+ = N(c)$ for $\sigma(c) = 1$; $N_- = N(c)$ for $\sigma(c) = -1$; and $N(0)$. We know N_+, as the solution count for Lehmers' equation (106):

$$N_+ = \begin{cases} p^2 - p + 1 + 4A_3(p)^2 & \text{if } p \equiv 1 \bmod 3, \\ p^2 - p + 3 & \text{if } p \equiv 2 \bmod 3. \end{cases}$$

We can easily evaluate $N(0)$, as

$$\#\left\{x, y, z \in \mathbb{F}_p^* : \frac{1}{x} + \frac{1}{y} + \frac{1}{z} = 0\right\} = \#\left\{x, y, z \in \mathbb{F}_p^* : x + y + z = 0\right\}$$

$$= \#\left\{x, y \in \mathbb{F}_p^* : x + y \ne 0\right\}$$

$$= (p - 1)(p - 2).$$

We are left with evaluating N_-. We use the overall count

$$\sum_{c \in \mathbb{F}_p} N(c) = (p - 1)^3 + (p - 1)n_0.$$

The right-hand side accounts for all the triples (x, y, z) with entries in $\mathbb{F}_p^*$, as well as for the left-hand side's overcount involving

$$n_0 = \# \left\{ x, y, z \in \mathbb{F}_p^* : \frac{1}{x} + \frac{1}{y} + \frac{1}{z} = x + y + z = 0 \right\}.$$

We note that n_0 was, in fact, computed in Step 1 of our discussion of Equation (106): $n_0 = (p - 1)(1 + \sigma(-3))$. Thus,

$$N(0) + \frac{p - 1}{2}(N_+ + N_-) = (p - 1)^3 + (p - 1)^2(1 + \sigma(-3)),$$

whence

$$N_+ + N_- = 2(p^2 - p + 1) + 2(p - 1)(\sigma(-3) - 1).$$

We deduce that

$$N_- = \begin{cases} p^2 - p + 1 - 4A_3(p)^2 & \text{if } p \equiv 1 \bmod 3, \\ p^2 - 5p + 3 & \text{if } p \equiv 2 \bmod 3. \end{cases}$$

Exercise 4.20. (i) To begin with,

$$\varphi_2(1)^2 = \sum_{x,y \in \mathbb{F}_p} \sigma(x^3 + x)\sigma(y^3 + y) = \sum_{x,y \in \mathbb{F}_p^*} \sigma\big(xy(x^2 + 1)(y^2 + 1)\big)$$

$$= \sum_{x,y \in \mathbb{F}_p^*} \sigma\big(x(x^2 y^2 + 1)(y^2 + 1)\big)$$

by making the change of variable $x := xy$ in the last step. As the argument is now a polynomial in y^2, we can use the downgrading formula:

$$\varphi_2(1)^2 = \sum_{x,y \in \mathbb{F}_p^*} \sigma\big(x(x^2 y + 1)(y + 1)\big) + \sum_{x,y \in \mathbb{F}_p^*} \sigma\big(xy(x^2 y + 1)(y + 1)\big).$$

In general, downgrading is done over $\mathbb{F}_p$ but it works over $\mathbb{F}_p^*$ just the same. Now, the first sum can be computed as follows:

$$\sum_{x,y \in \mathbb{F}_p^*} \sigma\big(x(x^2 y + 1)(y + 1)\big) = \sum_{x \in \mathbb{F}_p^*} \sigma(x) \sum_{y \in \mathbb{F}_p^*} \sigma\big((y + x^{-2})(y + 1)\big)$$

$$= \sum_{x \in \mathbb{F}_p^*} \sigma(x)\Big(p[\![x^{-2} = 1]\!] - 1 - \sigma(x^{-2})\Big)$$

$$= p\big(\sigma(1) + \sigma(-1)\big).$$

The second sum can be rewritten by the change of variable $y := yx^{-1}$:

$$\sum_{x,y\in\mathbb{F}_p^*} \sigma\big(xy(x^2y+1)(y+1)\big) = \sum_{x,y\in\mathbb{F}_p^*} \sigma\big(y(xy+1)(yx^{-1}+1)\big)$$

$$= \sum_{x,y\in\mathbb{F}_p^*} \sigma\big(xy(xy+1)(x+y)\big).$$

Clearly, the latter summation can be extended to $x,y \in \mathbb{F}_p$. All in all, we have

$$\sum_{x,y\in\mathbb{F}_p} \sigma\big(xy(xy+1)(x+y)\big) = \varphi_2(1)^2 - p\big(1+\sigma(-1)\big)$$

which equals 0 if $p \not\equiv 1 \bmod 4$, respectively $4A_2(p)^2 - 2p$ if $p \equiv 1 \bmod 4$.

(ii) The number of solutions to the equation $z^2 = (x^2 + y^2)(1 + x^2y^2)$ is

$$N = \sum_{x,y\in\mathbb{F}_p} \Big(1 + \sigma\big((1+x^2y^2)(x^2+y^2)\big)\Big)$$

$$= p^2 + \sum_{x,y\in\mathbb{F}_p} \sigma\big((1+x^2y^2)(x^2+y^2)\big).$$

Using the downgrading formula in y, we can write the latter sum as $S_1 + S_2$, where

$$S_1 = \sum_{x,y\in\mathbb{F}_p} \sigma\big((1+x^2y)(x^2+y)\big), \quad S_2 = \sum_{x,y\in\mathbb{F}_p} \sigma\big(y(1+x^2y)(x^2+y)\big).$$

We evaluate S_1 as follows:

$$S_1 = \sum_{x\in\mathbb{F}_p^*}\sum_{y\in\mathbb{F}_p} \sigma\big((y+x^{-2})(y+x^2)\big) = \sum_{x\in\mathbb{F}_p^*} \Big(p[\![x^{-2} = x^2]\!] - 1\Big)$$

$$= \big(3+\sigma(-1)\big)p - (p-1) = \big(2+\sigma(-1)\big)p + 1.$$

Along the way, we have used the fact that $x^{-2} = x^2$ has $3+\sigma(-1)$ solutions: two solutions from $x^2 = 1$, and $1 + \sigma(-1)$ solutions from $x^2 = -1$.

Next, we evaluate S_2 by using the downgrading formula in x:

$$S_2 = \sum_{x,y\in\mathbb{F}_p} \sigma\big(xy(1+xy)(x+y)\big) + \sum_{x,y\in\mathbb{F}_p} \sigma\big(y(1+xy)(x+y)\big).$$

The first sum has just been evaluated in part (i). The second sum is easily settled:

$$\sum_{x,y\in\mathbb{F}_p} \sigma\big(y(1+xy)(x+y)\big) = \sum_{y\in\mathbb{F}_p^*}\sum_{x\in\mathbb{F}_p} \sigma\big((x+y^{-1})(x+y)\big)$$

$$= \sum_{y\in\mathbb{F}_p^*}\Big(p[\![y^{-1}=y]\!] - 1\Big)$$

$$= 2p - (p-1) = p+1.$$

Thus,

$$S_2 = \varphi_2(1)^2 - p\big(1+\sigma(-1)\big) + p + 1 = \varphi_2(1)^2 - \sigma(-1)p + 1.$$

Summarizing, the solution count is

$$N = p^2 + S_1 + S_2 = (p+1)^2 + \varphi_2(1)^2 + 1.$$

Exercise 4.21. (i) We use the *xy-to-uv* change of variables that was so effective in solving Lehmers' equation. The correspondence is

$$u = \frac{x+1}{y}, \qquad v = \frac{y+1}{x},$$

with inverse

$$x = \frac{u+1}{uv-1}, \qquad y = \frac{v+1}{uv-1},$$

over domains $\{(u,v)\in\mathbb{F}_p^2 : u,v\neq -1, uv\neq 1\}$, respectively $\{(x,y)\in\mathbb{F}_p^2 : x,y\neq 0, x+y+1\neq 0\}$. On these domains, we have

$$\sum_{\substack{x,y\neq 0 \\ x+y+1\neq 0}} \sigma\left(\frac{(x+1)(y+1)(x-y)}{xy}\right) = \sum_{\substack{u,v\neq -1 \\ uv\neq 1}} \sigma\left(\frac{uv(u-v)}{uv-1}\right). \tag{154}$$

Let us look at these sums over the excluded regions. On the one hand

$$\sum_{\substack{u=-1,v\neq -1 \\ uv\neq 1}} \sigma\left(\frac{uv(u-v)}{uv-1}\right) = \sum_{v\neq -1} \sigma(-v) = -\sigma(1) = -1.$$

A similar calculation over $v = -1, u \neq -1$ yields $-\sigma(-1)$. It follows that

$$\sum_{uv \neq 1} \sigma\left(\frac{uv(u-v)}{uv-1}\right) = -(1 + \sigma(-1)) + \sum_{\substack{u,v \neq -1 \\ uv \neq 1}} \sigma\left(\frac{uv(u-v)}{uv-1}\right).$$

On the other hand

$$\sum_{\substack{x,y \neq 0 \\ x+y+1=0}} \sigma\left(\frac{(x+1)(y+1)(x-y)}{xy}\right) = \sum_{x \neq 0,-1} \sigma(2x+1) = -(1 + \sigma(-1)).$$

We may thus upgrade (154) to

$$\sum_{x,y \neq 0} \sigma\left(\frac{(x+1)(y+1)(x-y)}{xy}\right) = \sum_{uv \neq 1} \sigma\left(\frac{uv(u-v)}{uv-1}\right).$$

Inverting the denominators, and then extending the summation range with harmless terms, we deduce that

$$\sum_{x,y \in \mathbb{F}_p} \sigma\big(xy(x+1)(y+1)(x-y)\big) = \sum_{u,v \in \mathbb{F}_p} \sigma\big(uv(uv-1)(u-v)\big)$$

$$= \sum_{u,v \in \mathbb{F}_p} \sigma\big(uv(uv+1)(u+v)\big).$$

The last step is simply the sign change $v := -v$.

(ii) The number of solutions to $w^2 = (x^2 - y^2)(y^2 - z^2)(z^2 - x^2)$ is

$$N = \sum_{x,y,z \in \mathbb{F}_p} \Big(1 + \sigma\big((x^2 - y^2)(y^2 - z^2)(z^2 - x^2)\big)\Big)$$

$$= p^3 + \sum_{x,y,z \in \mathbb{F}_p} \sigma\big((x^2 - y^2)(y^2 - z^2)(z^2 - x^2)\big).$$

Let Σ denote the latter sum. Firstly, we use the homogeneity of the polynomial argument so as to pass from a triple sum to a double sum.

The contribution of $z = 0$ to Σ is

$$\sum_{x,y\in\mathbb{F}_p} \sigma\big(-(x^2-y^2)y^2x^2\big) = \sum_{x\in\mathbb{F}_p^*}\sum_{y\in\mathbb{F}_p^*} \sigma(y^2-x^2) = \sum_{x\in\mathbb{F}_p^*} \big(-1-\sigma(-x^2)\big)$$

$$= -\big(1+\sigma(-1)\big)(p-1).$$

For $z \neq 0$, we rescale the first two variables $x := xz$, $y := yz$. We are led to

$$\Sigma = -\big(1+\sigma(-1)\big)(p-1) + (p-1)\sum_{x,y\in\mathbb{F}_p} \sigma\big((x^2-y^2)(y^2-1)(1-x^2)\big).$$

The argument in the latter double sum is polynomial in x^2 and y^2, so we can use the downgrading formula in each variable. Firstly, we downgrade with respect to y. The above double sum can thus be written as $S_1 + S_2$, where

$$S_1 = \sum_{x,y\in\mathbb{F}_p} \sigma\big((x^2-y)(y-1)(1-x^2)\big)$$

$$S_2 = \sum_{x,y\in\mathbb{F}_p} \sigma\big(y(x^2-y)(y-1)(1-x^2)\big).$$

Now,

$$S_1 = \sum_{x\in\mathbb{F}_p} \sigma(x^2-1)\sum_{y\in\mathbb{F}_p} \sigma\big((y-x^2)(y-1)\big)$$

$$= \sum_{x\in\mathbb{F}_p} \sigma(x^2-1)\big(p[\![x^2=1]\!]-1\big) = 1.$$

Thus far

$$\Sigma = -\sigma(-1)(p-1) + (p-1)S_2,$$

and we turn to evaluating S_2. Downgrading with respect to x, we write $S_2 = T_1 + T_2$ where

$$T_1 = \sum_{x,y\in\mathbb{F}_p} \sigma\big(y(x-y)(y-1)(1-x)\big)$$

$$T_2 = \sum_{x,y\in\mathbb{F}_p} \sigma\big(xy(x-y)(y-1)(1-x)\big).$$

Here, once again, we can quickly evaluate the first sum as

$$T_1 = \sum_{y \in \mathbb{F}_p} \sigma\big(-y(y-1)\big) \sum_{x \in \mathbb{F}_p} \sigma\big((x-y)(x-1)\big)$$

$$= \sum_{y \in \mathbb{F}_p} \sigma\big(-y(y-1)\big)\big(p[\![y=1]\!] - 1\big) = \sigma(-1).$$

Therefore,

$$\Sigma = (p-1)T_2 = (p-1) \sum_{x,y \in \mathbb{F}_p} \sigma\big(xy(x+1)(y+1)(x-y)\big)$$

after effecting a double sign change $x := -x$, $y := -y$ in T_2. We may now use part (i), as well as part (i) of the previous exercise, to conclude that the desired number of solutions is

$$N = p^3 + \Sigma = \begin{cases} p^3 + (p-1)\big(4A_2(p)^2 - 2p\big) & \text{if } p \equiv 1 \bmod 4, \\ \\ p^3 & \text{if } p \not\equiv 1 \bmod 4. \end{cases}$$

Chapter 5

Exercise 5.24. The first identity is immediate:

$$\sum_{c \in \mathbb{F}_p} \varrho(c) = \sum_{x \in \mathbb{F}_p} \sigma(x) \sum_{c \in \mathbb{F}_p} \sigma(c + x + x^2) = 0,$$

by using (5). For the second identity, we write

$$\sum_{c \in \mathbb{F}_p} \varrho(c)^2 = \sum_{x,y \in \mathbb{F}_p} \sigma(xy) \sum_{c \in \mathbb{F}_p} \sigma(c + x + x^2)\sigma(c + y + y^2).$$

For each $x, y \in \mathbb{F}_p$, the inner sum evaluates to $p[\![x + x^2 = y + y^2]\!] - 1$. Since $\sum_{x,y \in \mathbb{F}_p} \sigma(xy) = 0$, we obtain

$$\sum_{c \in \mathbb{F}_p} \varrho(c)^2 = p \sum_{\substack{x,y \in \mathbb{F}_p \\ x+x^2=y+y^2}} \sigma(xy).$$

Now, $x + x^2 = y + y^2$ if and only if $x = y$, or $x + y = -1$. We split the above sum accordingly: the part subjected to the condition $x = y$ equals $p - 1$, while the part subjected to the condition $x + y = -1$ equals $-\sigma(-1)$.

The two sums have one term in common, corresponding to $x = y = -1/2$; that term equals 1. Hence,

$$\sum_{c \in \mathbb{F}_p} \varrho(c)^2 = p\big(p - 2 - \sigma(-1)\big).$$

Exercise 5.25. Set

$$S(a) = \sum_{c \in \mathbb{F}_p} \varrho(c)\varrho(ac).$$

We note that $S(0) = \varrho(0)\sum_{c \in \mathbb{F}_p} \varrho(c) = 0$; also $S(1) = \sum_{c \in \mathbb{F}_p} \varrho(c)^2$ was already computed in the previous exercise. Assume $a \neq 0, 1$ in what follows. Following the same steps as above, we have

$$S(a) = \sum_{x,y \in \mathbb{F}_p} \sigma(xy) \sum_{c \in \mathbb{F}_p} \sigma(c + x + x^2)\sigma(ac + y + y^2)$$

$$= \sum_{x,y \in \mathbb{F}_p} \sigma(xy)\sigma(a)\Big(p[\![a(x + x^2) = y + y^2]\!] - 1\Big) = p\sigma(a)\Sigma,$$

where

$$\Sigma = \sum_{\substack{x,y \in \mathbb{F}_p \\ a(x+x^2)=y+y^2}} \sigma(xy).$$

We may discard the value $x = 0$. For $x \neq 0$, we change the variable $y = tx$. Then $\sigma(xy) = \sigma(t)$, and the constraint $a(x + x^2) = y + y^2$ amounts to $ax + ax^2 = tx + t^2x^2$, or $a + ax = t + t^2x$. Therefore,

$$\Sigma = \sum_{t \in \mathbb{F}_p} \sigma(t) \cdot \#\{x \neq 0 : a + ax = t + t^2x\}.$$

Recall that $a^2 \neq a$. The equation $a + ax = t + t^2x$ has no non-zero solutions if $t = a$; no solutions if $t^2 = a$; one non-zero solution if $t \neq a$ and $t^2 \neq a$.

Thus,

$$\Sigma = \sum_{t \neq a,\, t^2 \neq a} \sigma(t).$$

When a is not a square, we get $\Sigma = -\sigma(a) = 1$; correspondingly,

$$S(a) = -p.$$

When a is a square, say $a = d^2$, we get $\Sigma = -\sigma(a) - \sigma(d) - \sigma(-d) = -1 - \sigma(d) - \sigma(-d)$; correspondingly,

$$S(a) = -\big(1 + \sigma(d) + \sigma(-d)\big)p.$$

In particular, for $p \equiv 3 \bmod 4$, we have $S(a) = -p$ for each $a \neq 0, 1$.

Next, set

$$T(a) = \sum_{c \in \mathbb{F}_p} \varrho(c)\varrho(a + c).$$

As $T(0) = \sum_{c \in \mathbb{F}_p} \varrho(c)^2$ is already known, we assume $a \neq 0$ in what follows. We write

$$T(a) = \sum_{x,y \in \mathbb{F}_p} \sigma(xy) \sum_{c \in \mathbb{F}_p} \sigma(c + x + x^2)\sigma(a + c + y + y^2)$$

$$= \sum_{x,y \in \mathbb{F}_p} \sigma(xy)\big(p[\![x + x^2 = a + y + y^2]\!] - 1\big) = p\Sigma',$$

where

$$\Sigma' = \sum_{\substack{x,y \in \mathbb{F}_p \\ x + x^2 = a + y + y^2}} \sigma(xy).$$

Again, we may discard the value $x = 0$, and for $x \neq 0$ we effect a change the variable $y = tx$. Then

$$\Sigma' = \sum_{t \in \mathbb{F}_p} \sigma(t) \cdot \#\{x \neq 0 : x + x^2 = a + tx + t^2 x^2\}.$$

Consider the equation

$$x + x^2 = a + tx + t^2 x^2,$$

viewed as a quadratic equation in x, and recall that $a \neq 0$. Then any solution automatically satisfies $x \neq 0$. Once rewritten as $(t^2 - 1)x^2 + (t - 1)x + a = 0$, we see that there are $1 + \sigma\big((t-1)^2 - 4a(t^2 - 1)\big)$ solutions; the only exceptions are the values $t = 1$, when there are no solutions, and $t = -1$, when there is one solution. Summarizing,

$$\Sigma' = \sigma(-1) + \sum_{t \neq \pm 1} \sigma(t)\left(1 + \sigma\left((t-1)^2 - 4a(t^2 - 1)\right)\right)$$

$$= -1 - \sigma(-1) + \sum_{t \in \mathbb{F}_p} \sigma(t)\left(1 + \sigma\left((t-1)^2 - 4a(t^2 - 1)\right)\right)$$

$$= -1 - \sigma(-1) + \sum_{t \in \mathbb{F}_p} \sigma(t)\sigma\left((1 - 4a)t^2 - 2t + (1 + 4a)\right).$$

Let Σ'' denote the latter sum. If $a = 1/4$, we easily get $\Sigma'' = -\sigma(-2)$. If $a \neq 1/4$ then, using (117), we get

$$\Sigma'' = \sigma(1 - 4a) \sum_{t \in \mathbb{F}_p} \sigma(t) \, \sigma\left(t^2 - \frac{2}{1 - 4a}t + \frac{1 + 4a}{1 - 4a}\right)$$

$$= \sigma(1 - 4a) \, \sigma\left(-\frac{2}{1 - 4a}\right) \varrho\left(\frac{(1 + 4a)(1 - 4a)}{4}\right)$$

$$= \sigma(-2)\varrho\left(\frac{1}{4} - 4a^2\right) = \varrho(4a^2).$$

In the last step, we have used the twisted symmetry. Note that the final formula, $\Sigma'' = \varrho(4a^2)$, is valid when $a = 1/4$ as well. We conclude that, for $a \neq 0$, we have

$$T(a) = \left(\varrho(4a^2) - 1 - \sigma(-1)\right)p.$$

Exercise 5.26. Consider the sum

$$S = \sum_{x \in \mathbb{F}_p} \left(1 - \sigma(x)\right)\left(1 - \sigma(x^2 + x + c)\right).$$

On the one hand, we can evaluate the sum S by expanding:

$$S = \sum_{x \in \mathbb{F}_p} 1 - \sum_{x \in \mathbb{F}_p} \sigma(x) - \sum_{x \in \mathbb{F}_p} \sigma(x^2 + x + c) + \sum_{x \in \mathbb{F}_p} \sigma(x)\sigma(x^2 + x + c)$$

$$= p + 1 + \varrho(c).$$

Since $p \equiv 2 - \sigma(-1) \bmod 4$, we deduce that

$$S \equiv \varrho(c) - 1 - \sigma(-1) \bmod 4.$$

On the other hand, each term in S is either 0 or 4, except when $x = 0$ or $x^2 + x + c = 0$. The term corresponding to $x = 0$ contributes $1 - \sigma(c)$ to S. When $\sigma(1 - 4c) = 1$, the two solutions x_1 and x_2 to $x^2 + x + c = 0$ contribute $(1 - \sigma(x_1)) + (1 - \sigma(x_2))$ to S. As $(1 - \sigma(x_1))(1 - \sigma(x_2)) \equiv 0$ mod 4, we deduce that

$$(1 - \sigma(x_1)) + (1 - \sigma(x_2)) \equiv 1 - \sigma(x_1 x_2) = 1 - \sigma(c) \text{ mod } 4.$$

Thus, modulo 4,

$$S \equiv \begin{cases} 1 - \sigma(c) & \text{if } \sigma(1 - 4c) = -1 \\ 2(1 - \sigma(c)) & \text{if } \sigma(1 - 4c) = 1 \end{cases}$$

$$\equiv \begin{cases} 2 & \text{if } \sigma(c) = \sigma(1 - 4c) = -1 \\ 0 & \text{otherwise.} \end{cases}$$

The claimed formula for $\varrho(c)$ mod 4 follows by combining the two evaluations of S modulo 4.

Exercise 5.27. We wish to check that

$$\sum_{x \in \mathbb{F}_p} \sigma(x)\sigma(a^2 + acx + dx^2)$$

$$= \sum_{x \in \mathbb{F}_p} \sigma(-x)\sigma(a^2 + 2acx + (c^2 - 4d)x^2). \qquad (*)$$

When $a = 0$, the identity $(*)$ is clearly true as both sides vanish. When $a \neq 0$, the change of variable $x := ax$ turns $(*)$ into

$$\sum_{x \in \mathbb{F}_p} \sigma(x)\sigma(1 + cx + dx^2) = \sum_{x \in \mathbb{F}_p} \sigma(-x)\sigma(1 + 2cx + (c^2 - 4d)x^2). \qquad (*_1)$$

Next, the change of variable $x := 1/x$ on $\mathbb{F}_p^*$ turns $(*_1)$ into

$$\sum_{x \in \mathbb{F}_p} \sigma(x)\sigma(x^2 + cx + d) = \sum_{x \in \mathbb{F}_p} \sigma(-x)\sigma(x^2 + 2cx + (c^2 - 4d)). \qquad (*_2)$$

When $c = 0$, the identity $(*_2)$ amounts to an equality involving Jacobsthal sums: $\varphi_2(d) = \sigma(-1)\varphi_2(-4d)$. Let us check it. Using (65) and periodicity, we have $\varphi_2(-4d) = \sigma(2)\varphi_2(4d) = \sigma(2)\sigma(2)\varphi_2(d) = \varphi_2(d)$. This agrees with $\sigma(-1)\varphi_2(d)$ when $p \equiv 1$ mod 4, since $\sigma(-1) = 1$, as well as when $p \equiv 3$ mod 4, for $\varphi_2(d) = 0$ in this case.

When $c \neq 0$, an appeal to (117) turns $(*_2)$ into an identity in terms of ϱ-sums:

$$\sigma(c)\varrho\left(\frac{d}{c^2}\right) = \sigma(-1)\sigma(2c)\varrho\left(\frac{c^2 - 4d}{4c^2}\right). \tag{$*_3$}$$

After cancelling $\sigma(c)$, it becomes evident that $(*_3)$ amounts to the twisted symmetry of ϱ-sums.

Exercise 5.28. We evaluate the first sum

$$\Sigma_1 = \sum_{x \in \mathbb{F}_p} \sigma(x^4 - 6x^2 - 16x + 41) \tag{155}$$

by resorting to Theorem 5.16. We wish to write the quartic argument in the form $(x^2 + b)^2 + c(x + d)^2$; the values $b = -5$, $c = 4$, $d = -2$ are quickly found. Then $r = -1$, $s = 24$, and we note that c, r, s are all non-zero. Thus,

$$\Sigma_1 = -1 + \sigma(6)\varrho\left(\frac{1}{36}\right) = -1 + \sigma(-2)\varphi_2(1)$$

$$= \begin{cases} 2A_2(p) - 1 & \text{if } p \equiv 1 \bmod 8, \\ -2A_2(p) - 1 & \text{if } p \equiv 5 \bmod 8, \\ -1 & \text{otherwise.} \end{cases}$$

We can write the second sum as follows:

$$\Sigma_2 = \sum_{x \in \mathbb{F}_p} \sigma(x^5 + 4x^4 + bx^3 + 4x^2 + x), \qquad b := \frac{38}{9}.$$

The argument is a palindromic quintic, just like the ones covered by Theorem 5.20. We are, however, in the case when $a = 4$. We follow along the proof of Theorem 5.20, adjusting as needed. We are working under the condition $b \neq -2a - 2$, that is $b \neq -10$. The change occurs at the point where we evaluate S_4, for now $B = 0$. Thus,

$$S_4 = -\sigma(\alpha) + \sigma(\alpha)\varphi_2\left(\frac{\beta^2 - 4\alpha\gamma}{\alpha^2}\right).$$

When $a = 4$ we have $2\alpha + \beta = 2$ and $\alpha - \gamma = 4$. One checks that $\beta^2 - 4\alpha\gamma = 4(1 - \alpha)$. By combining simplified the expressions of S_3 and S_4, we reach the formula

$$\Sigma_2 = \sigma(2)\varrho\left(\frac{\alpha}{4}\right) + \sigma(\alpha)\varphi_2\left(\frac{4(1-\alpha)}{\alpha^2}\right).$$

At this point we use the value $b = 38/9$ to compute $\alpha = (b+10)/16 = 8/9$. Thus

$$\Sigma_2 = \sigma(2)\varrho\left(\frac{2}{9}\right) + \sigma(2)\varphi_2\left(\frac{9}{16}\right).$$

Moreover, $\varrho(2/9) = \sigma(6)\varphi_2(1)$ and $\varphi_2(9/16) = \sigma(3/4)\varphi_2(1) = \sigma(3)\varphi_2(1)$. We conclude that

$$\Sigma_2 = (1 + \sigma(2))\sigma(3)\varphi_2(1) = \begin{cases} 4A_2(p) & \text{if } p \equiv 1 \bmod 24, \\ -4A_2(p) & \text{if } p \equiv 17 \bmod 24, \\ 0 & \text{otherwise.} \end{cases}$$

In the third sum

$$\Sigma_3 = \sum_{x \in \mathbb{F}_p} \sigma(x^6 - 15x^4 + 15x^2 - 1) \tag{156}$$

the sextic argument factors as $(x^2 - 1)(x^4 - 14x^2 + 1)$. This being a polynomial in x^2, we can use the downgrading formula to write

$$\Sigma_3 = \sum_{x \in \mathbb{F}_p} \sigma\big((x-1)(x^2 - 14x + 1)\big) + \sum_{x \in \mathbb{F}_p} \sigma\big(x(x-1)(x^2 - 14x + 1)\big).$$

We evaluate the first sum by changing the variable $x := x + 1$, followed by an appeal to (117). The outcome is $\sigma(-3)\varrho(-1/12) = \sigma(6)\psi_3(1)$. We evaluate the second sum by applying Theorem 5.10. The outcome is $-1 + \sigma(-6)\varrho(1/3) = -1 + \sigma(-6)\psi_3(1)$. All in all,

$$\Sigma_3 = -1 + \big(\sigma(6) + \sigma(-6)\big)\psi_3(1) = \begin{cases} 4A_3(p) - 1 & \text{if } p \equiv 1 \bmod 24, \\ -4A_3(p) - 1 & \text{if } p \equiv 13 \bmod 24, \\ -1 & \text{otherwise.} \end{cases}$$

Exercise 5.29. (i) Firstly, the change of variable $x := -x$ gives

$$\sum_{x \in \mathbb{F}_p} \sigma(x^5 + bcx^3 + c^2 x) = \sigma(-1) \sum_{x \in \mathbb{F}_p} \sigma(x^5 + bcx^3 + c^2 x)$$

so the left-hand sum vanishes unless $p \equiv 1 \bmod 4$. Secondly, the change of variable $x := cx^{-1}$ on $\mathbb{F}_p^*$ turns each term $\sigma(x^5 + bcx^3 + c^2 x)$ into

$$\sigma(c^5 x^{-5} + bc^4 x^{-3} + c^3 x^{-1}) = \sigma(c^5 x + bc^4 x^3 + c^3 x^5)$$
$$= \sigma(c)\sigma(x^5 + bcx^3 + c^2 x).$$

It follows that

$$\sum_{x \in \mathbb{F}_p} \sigma(x^5 + bcx^3 + c^2 x) = \sigma(c) \sum_{x \in \mathbb{F}_p} \sigma(x^5 + bcx^3 + c^2 x)$$

so the left-hand sum vanishes unless $\sigma(c) = 1$.

Assume now that $p \equiv 1 \bmod 4$ and $\sigma(c) = 1$. Put $c = t^2$ for some $t \in \mathbb{F}_p^*$. The change of variable $x := tx$ on $\mathbb{F}_p$ turns each term $\sigma(x^5 + bcx^3 + c^2 x) = \sigma(x^5 + bt^2 x^3 + t^4 x)$ into $\sigma(t^5 x^5 + bt^5 x^3 + t^5 x) = \sigma(t)\sigma(x^5 + bx^3 + x)$. By Corollary 5.22, we obtain

$$\sum_{x \in \mathbb{F}_p} \sigma(x^5 + bcx^3 + c^2 x) = \sigma(t) \sum_{x \in \mathbb{F}_p} \sigma(x^5 + bx^3 + x) = 2\sigma(t)\varrho\left(\frac{b+2}{16}\right).$$

The desired computation is thereby checked; furthermore, we also know how to determine the sign.

(ii) We use the findings from part (i), where now $c = 3a$ and $b = 10/3$:

$$\sum_{x \in \mathbb{F}_p} \sigma(x^5 + 10ax^3 + 9a^2 x) = \begin{cases} 2\sigma(t)\varrho\left(\dfrac{1}{3}\right) & \text{if } p \equiv 1 \bmod 4,\ \sigma(3a) = 1, \\ 0 & \text{otherwise,} \end{cases}$$

where, in the first case, t is a square root of $3a$ in $\mathbb{F}_p$. We have $\varrho(1/3) = \psi_3(1) = 2A_3(p)$ if $p \equiv 1 \bmod 3$, and 0 otherwise. Thus,

$$\sum_{x \in \mathbb{F}_p} \sigma(x^5 + 10ax^3 + 9a^2 x) = \begin{cases} \sigma(t) \cdot 4A_3(p) & \text{if } p \equiv 1 \bmod 12,\ \sigma(3a) = 1, \\ 0 & \text{otherwise.} \end{cases}$$

Moreover, for $p \equiv 1 \bmod 12$ we have $\sigma(3) = 1$; thus $\sigma(3a) = 1$ is equivalent to $\sigma(a) = 1$. The above computation yields $\pm 4A_3(p)$ when $p \equiv 1 \bmod 12$ and $\sigma(a) = 1$, and 0 otherwise. This proves the main claim.

Consider now the case $a = -1$, still assuming that $p \equiv 1 \bmod 12$. Thus, $\sigma(a) = -1$. The quadratic sum evaluates to $\sigma(t) \cdot 4A_3(p)$, where t is a square root of -3 in $\mathbb{F}_p$. If we write $p = A_3(p)^2 + 3B^2$ then we can take

$t = A_3(p)/B \in \mathbb{F}_p$. We get $\sigma(t) = \sigma\big(A_3(p)B\big)$ and so

$$\sum_{x \in \mathbb{F}_p} \sigma(x^5 - 10x^3 + 9x) = \sigma\big(A_3(p)B\big) \cdot 4A_3(p).$$

Exercise 5.30. Put

$$S(a) = \sum_{x \in \mathbb{F}_p} \sigma(x^7 + ax^5 + ax^3 + x). \tag{157}$$

The septic argument factors as follows:

$$x(x^6 + ax^4 + ax^2 + 1) = x(x^2 + 1)\big(x^4 + (a - 1)x^2 + 1\big)$$
$$= x(x^2 + 1)\big((x^2 + 1)^2 + (a - 3)x^2\big).$$

Therefore,

$$S(a) = \sum_{x \in \mathbb{F}_p} \sigma\Big(x(x^2 + 1)\big((x^2 + 1)^2 + (a - 3)x^2\big)\Big)$$

$$= \sum_{x \neq 0} \sigma\left(\frac{x^2 + 1}{x} \cdot \left(\left(\frac{x^2 + 1}{x}\right)^2 + a - 3\right)\right)$$

$$= \sum_{y \in \mathbb{F}_p} \sigma\big(y(y^2 + a - 3)\big) \cdot \#\left\{x \in \mathbb{F}_p^* : y = \frac{x^2 + 1}{x}\right\}$$

$$= \sum_{y \in \mathbb{F}_p} \sigma\big(y(y^2 + a - 3)\big) \cdot \big(1 + \sigma(y^2 - 4)\big)$$

$$= \varphi_2(a - 3) + \Sigma(a),$$

where

$$\Sigma(a) = \sum_{y \in \mathbb{F}_p} \sigma\big(y(y^2 + a - 3)(y^2 - 4)\big).$$

Note that we have descended from a septic argument, in $S(a)$, to a quintic argument in $\Sigma(a)$. It is the computation of $\Sigma(a)$ that uses the specified parameter values.

Let $a = 7$. We then get

$$\Sigma(7) = \sum_{y \in \mathbb{F}_p} \sigma\big(y(y^2 + 4)(y^2 - 4)\big) = \sum_{y \in \mathbb{F}_p} \sigma\big(y(y^4 - 16)\big) = \varphi_4(-16).$$

Thus,

$$S(7) = \varphi_2(4) + \varphi_4(-16) = \sigma(2)\varphi_2(1) + \sigma(2)\varphi_4(-1).$$

Recall that $\varphi_4(-1) = 0$ if $p \not\equiv 1 \bmod 8$, and $\varphi_4(-1) = (-1)^{(p-1)/8}\varphi_4(1)$ if $p \equiv 1 \bmod 8$. Using the known evaluations of $\sigma(2)$, $\varphi_2(1)$, and $\varphi_4(1)$, we deduce that

$$S(7) = \begin{cases} 2A_2(p) + 4A_4(p) & \text{if } p \equiv 1 \bmod 8, \\ -2A_2(p) & \text{if } p \equiv 5 \bmod 8, \\ 0 & \text{if } p \equiv 3, 7 \bmod 8. \end{cases}$$

Let $a = -33$. We have

$$\Sigma(-33) = \sum_{y \in \mathbb{F}_p} \sigma\big(y(y^2 - 36)(y^2 - 4)\big) = \sigma(2) \sum_{y \in \mathbb{F}_p} \sigma\big(y(y^2 - 9)(y^2 - 1)\big)$$

after a rescaling $y := 2y$. The latter sum is precisely the one computed in part (ii) of the previous exercise, as long as $p > 3$. We thus get

$$\Sigma(-33) = \begin{cases} \sigma\big(2A_3(p)B\big) \cdot 4A_3(p) & \text{if } p \equiv 1 \bmod 12, \\ 0 & \text{otherwise,} \end{cases}$$

where, we recall, $p = A_3(p)^2 + 3B^2$. Now,

$$S(-33) = \varphi_2(-36) + \Sigma(-33)$$

and we note that $\varphi_2(-36) = \sigma(6)\varphi_2(-1) = \sigma(3)\varphi_2(1)$, the latter by (65). All in all

$$S(-33) = \begin{cases} 2A_2(p) + \sigma\big(2A_3(p)B\big) \cdot 4A_3(p) & \text{if } p \equiv 1 \bmod 12, \\ -2A_2(p) & \text{if } p \equiv 5 \bmod 12, \\ 0 & \text{if } p \equiv 7, 11 \bmod 12. \end{cases}$$

Exercise 5.31. The equation

$$(x + y + z)^3 = 54xyz \tag{158}$$

has p solutions in which $z = 0$. When $z \neq 0$, the change of variables $x := xz$, $y := yz$ turns the homogeneous Equation (158) into the inhomogeneous

$$(x + y + 1)^3 = 54xy. \tag{$*$}$$

The number of solutions to (158) is thus $p + (p-1)N_*$, where N_* is the number of solutions to $(*)$.

Next, the shift $x := x - (y + 1)$ turns $(*)$ into $x^3 = 54(x - y - 1)y$, which we write as $54y^2 - 54(x-1)y + x^3 = 0$. This is a quadratic equation in y; for each $x \in \mathbb{F}_p$ there are $1 + \sigma(\Delta(x))$ solutions, where

$$\Delta(x) = \big(-54(x-1)\big)^2 - 4 \cdot 54 \cdot x^3 = -54\big(4x^3 - 54(x-1)^2\big).$$

Thus,

$$N_* = \sum_{x \in \mathbb{F}_p} \big(1 + \sigma(\Delta(x))\big) = p + \sigma(-54) \sum_{x \in \mathbb{F}_p} \sigma\big(4x^3 - 54(x-1)^2\big)$$

$$= p + \sigma(-1) \sum_{x \in \mathbb{F}_p} \sigma\big(x^3 - (3x-2)^2\big),$$

where, in the last step, we rescaled $x := 3x/2$ in the sum. Visibly, the cubic argument $x^3 - (3x-2)^2$ is reducible, as it admits the root $x = 1$. The shift $x := x + 1$ yields

$$\sum_{x \in \mathbb{F}_p} \sigma\big(x^3 - (3x-2)^2\big) = \sum_{x \in \mathbb{F}_p} \sigma(x^3 - 6x^2 - 3x) = \sigma(-6)\varrho\left(-\frac{1}{12}\right)$$

$$= \sigma(-6) \cdot \sigma(-2)\psi_3(1) = \sigma(3)\psi_3(1).$$

Summarizing, the number of solutions to (158) is

$$p + (p-1)\big(p + \sigma(-3)\psi_3(1)\big) = p^2 + \sigma(-3)\psi_3(1)(p-1)$$

$$= \begin{cases} p^2 + 2A_3(p)(p-1) & \text{if } p \equiv 1 \bmod 3, \\ p^2 & \text{if } p \equiv 2 \bmod 3. \end{cases}$$

Exercise 5.32. Let $a \in \mathbb{F}_p^*$, and consider the equation

$$(x + y + z + t + 1)^2 = axyzt. \tag{159}$$

Following the same steps as in Exercise 2.14, we view (159) as a quadratic equation in t; it has $1 + \sigma(axyz)\sigma(axyz - 4(x + y + z + 1))$ solutions for each $x, y, z \in \mathbb{F}_p$. Hence, the number of solutions to the Equation (159) is

$$N(a) = \sum_{x,y,z \in \mathbb{F}_p} \left(1 + \sigma(axyz)\sigma(axyz - 4x - 4y - 4z - 4)\right)$$

$$= p^3 + \sum_{x,y \in \mathbb{F}_p} \sigma(axy) \sum_{z \in \mathbb{F}_p} \sigma(z)\sigma((axy - 4)z - 4(x + y + 1)).$$

We can evaluate the inner-most sum by using (150). We obtain

$$N(a) = p^3 + \sum_{x,y \in \mathbb{F}_p} \sigma(axy)\sigma(axy - 4)\left(\llbracket x + y + 1 = 0 \rrbracket p - 1\right)$$

$$= p^3 + \sum_{x,y \in \mathbb{F}_p} \sigma(axy)\sigma(axy - 4) + p \sum_{x+y+1=0} \sigma(axy)\sigma(axy - 4)$$

Next, we evaluate the two sums. The first sum is straightforward:

$$\sum_{x,y \in \mathbb{F}_p} \sigma(axy)\sigma(axy - 4) = \sum_{x \in \mathbb{F}_p} \sigma(ax) \sum_{y \in \mathbb{F}_p} \sigma\big(y(axy - 4)\big)$$

$$= \sum_{x \in \mathbb{F}_p} \sigma(ax) \cdot (-\sigma(ax)) = -(p - 1).$$

It is the evaluation of the second sum, which we denote by Σ, that brings forth a ϱ-sum. Turning Σ into a single-variable sum, we write

$$\Sigma = \sum_{x \in \mathbb{F}_p} \sigma\big(-ax(x + 1)\big)\sigma\big(-ax(x + 1) - 4\big)$$

$$= \sum_{x \in \mathbb{F}_p} \sigma\big((x^2 + x)(x^2 + x + 4a^{-1})\big).$$

Now, we apply Theorem 5.10. As $\Delta_1 = 1$, $\Delta_2 = 1 - 16a^{-1}$, $B = 16a^{-1} - 2$, we find

$$\Sigma = \begin{cases} -1 + \sigma(16a^{-1} - 2)\varrho\left(\dfrac{1 - 16a^{-1}}{(2 - 16a^{-1})^2}\right) & \text{if } a \neq 8, \\[2em] -1 + \varphi_2(-1) & \text{if } a = 8. \end{cases}$$

All in all, the number of solutions to the Equation (159) is

$$N(a) = \begin{cases} p^3 - 1 + \sigma(16a^{-1} - 2)\varrho\left(\dfrac{1 - 16a^{-1}}{(2 - 16a^{-1})^2}\right)p & \text{if } a \neq 8, \\[2em] p^3 - 1 + \varphi_2(-1)p & \text{if } a = 8. \end{cases}$$

We note that the count is explicit when $a = 8$. When $a \neq 8$, it can still be made explicit for a handful of parameter values. For $a = 32$, which is the original question, we obtain

$$N(32) = p^3 - 1 + \sigma(-6)\varrho\left(\frac{2}{9}\right)p = p^3 - 1 + \sigma(-1)\varphi_2(1)p$$

$$= \begin{cases} p^3 - 1 + 2A_2(p)p & \text{if } p \equiv 1 \bmod 4, \\ p^3 - 1 & \text{if } p \equiv 3 \bmod 4. \end{cases}$$

For $a = -16$, we have

$$N(-16) = p^3 - 1 + \sigma(-3)\varrho\left(\frac{2}{9}\right)p = p^3 - 1 + \sigma(-2)\varphi_2(1)p$$

which happens to equal $N(32)$. Finally, for $a = 16$, we have

$$N(16) = p^3 - 1 + \sigma(-1)\varrho(0) = p^3 - 1 - \sigma(-1).$$

Exercise 5.33. For each $c \in \mathbb{F}_p$, consider the equations

$$x + \frac{1}{x} = 2b - c, \quad y + \frac{1}{y} = 2b + c.$$

These are quadratic equations in x respectively y, with discriminants $c^2 \pm 4bc + 4b^2 - 4$. It follows that

$$N(b) = \sum_{c \in \mathbb{F}_p} \left(1 + \sigma(c^2 - 4bc + 4b^2 - 4)\right)\left(1 + \sigma(c^2 + 4bc + 4b^2 - 4)\right)$$

$$= p - 2 + \sum_{c \in \mathbb{F}_p} \sigma\left((c^2 - 4bc + 4b^2 - 4)(c^2 + 4bc + 4b^2 - 4)\right).$$

We evaluate the latter sum by means of Theorem 5.10. As $\Delta_1 = \Delta_2 = 16$ and $B = 16(4b^2 - 2)$, we obtain

$$N(b) = p - 3 + \sigma(4b^2 - 2)\varrho\left(\left(\frac{1}{4b^2 - 2}\right)^2\right) \tag{160}$$

as long as $b^2 \neq 1/2$; when $b^2 = 1/2$, we have $N(b) = p - 3 + \varphi_2(16^2) = p - 3 + \varphi_2(1)$.

We turn to checking that $N(1/b) = N(b)$. The case $p = 3$ is trivial, as $1/b = b$ for any $b \in \mathbb{F}_p^*$; assume $p > 3$ in what follows.

If $b^2 = 1/2$ then $(1/b)^2 = 2$, so

$$N\left(\frac{1}{b}\right) = p - 3 + \sigma(6)\varrho\left(\frac{1}{36}\right) = p - 3 + \sigma(-2)\varphi_2(1).$$

Note that $\sigma(2) = 1$, as 2 is a square, and that $\sigma(-1)\varphi_2(1) = \varphi_2(1)$. Indeed, this holds when $p \equiv 3 \bmod 4$, since $\varphi_2(1) = 0$, and it also holds when $p \equiv 1 \bmod 4$, as $\sigma(-1) = 1$. We conclude that $N(1/b) = N(b)$ in this case. By symmetry, $N(1/b) = N(b)$ also holds when $b^2 = 2$.

Assume now that $b^2 \neq 2, 1/2$. Set

$$c = \frac{1}{4b^2 - 2},$$

so $c \neq 1/6$. We evaluate

$$\eta(c) = \frac{2c + 1}{12c - 2} = \frac{4b^2}{16 - 8b^2} = \frac{1}{4(1/b)^2 - 2}.$$

We thus have, by (160), that

$$\begin{cases} N(b) & = (p - 3) - \sigma(c)\varrho(c^2), \\ N(1/b) & = (p - 3) - \sigma(\eta(c))\varrho(\eta(c)^2). \end{cases}$$

Thanks to Theorem 5.8, we can write

$$\sigma(\eta(c))\varrho(\eta(c)^2) = \sigma(\eta(c))\sigma(12c - 2)\varrho(c^2) = \sigma(2c + 1)\varrho(c^2) = \sigma(c)\varrho(c^2),$$

the latter equality owing to the observation that $2c + 1 = (4b^2)c$. We conclude that $N(1/b) = N(b)$ in this case as well.

Bibliography

S. Ahlgren, K. Ono, D. Penniston. Zeta functions of an infinite family of $K3$ surfaces, *Amer. J. Math.*, 124 (2002), no. 2, 353–368.

J. Ax. Zeroes of polynomials over finite fields, *Amer. J. Math.*, 86 (1964), 255–261.

J.M. Basilla. On the solution of $x^2 + dy^2 = m$, *Proc. Japan Acad. Ser. A Math. Sci.*, 80 (2004), no. 5, 40–41.

B.C. Berndt, R.J. Evans, K.S. Williams. *Gauss and Jacobi Sums*, Canadian Mathematical Society Series of Monographs and Advanced Texts, John Wiley & Sons, 1998.

B.W. Brewer. On certain character sums, *Trans. Amer. Math. Soc.*, 99 (1961), 241–245.

J. Brillhart. Note on representing a prime as a sum of two squares, *Math. Comp.*, 26 (1972), 1011–1013.

T. Budzinski, G. Lucchini Arteche. Un résultat de recherche obtenu grâce à des lycéens, *Gaz. Math.*, 151 (2017), 36–41.

L. Carlitz. Certain special equations in a finite field, *Monatsh. Math.*, 58 (1954), 5–12.

L. Carlitz. The number of solutions of some equations in a finite field, *Portugal. Math.*, 13 (1954), 25–31.

A. Cerbu, E. Gunther, M. Magee, L. Peilen. The cycle structure of a Markoff automorphism over finite fields, *J. Number Theory*, 211 (2020), 1–27.

H.H. Chan, L. Long, Y. Yang. A cubic analogue of the Jacobsthal identity, *Amer. Math. Monthly*, 118 (2011), no. 4, 316–326.

S. Chowla. The last entry in Gauss's diary, *Proc. Nat. Acad. Sci. U.S.A.*, 35 (1940), 244–246.

S. Chowla. A formula similar to Jacobsthal's for the explicit value of x in $p = x^2 + y^2$ where p is a prime of the form $4k + 1$, *Proc. Lahore Philos. Soc.*, 7 (1945), 2 pp.

S. Chowla, J. Cowles, M. Cowles. On the number of zeros of diagonal cubic forms, *J. Number Theory* 9 (1977), no. 4, 502–506.

S. Chowla, J. Cowles, M. Cowles. The number of zeroes of $x^3 + y^3 + cz^3$ in certain finite fields, *J. Reine Angew. Math.*, 299/300 (1978), 406–410.

S. Chowla, J. Cowles, M. Cowles. On the difference of cubes (mod p), *Acta Arith.*, 37 (1980), 61–65.

D.A. Cox. *Primes of the Form x^2+ny^2. Fermat, Class Field Theory, and Complex Multiplication*, Second edition, John Wiley & Sons 2013.

R. Daublebsky von Sterneck. Über die Anzahl inkongruenter Werte, die eine ganze Funktion dritten Grades annimmt, Sitzungsber. *Akad. Wiss. Wien*, 116 (1907), 895–904.

H. Davenport. On certain exponential sums, *J. Reine Angew. Math.*, 169 (1933), 158–176.

M. de Courcy-Ireland. Non-planarity of Markoff graphs mod p, *Comment. Math. Helv.*, 99 (2024), no. 1, 111–148.

R.J. Evans. Determinations of Jacobsthal sums, *Pacific J. Math.* 110 (1984), no. 1, 49–58.

R.J. Evans. Identities for products of Gauss sums over finite fields, *Enseign. Math.*, 27 (1981), no. 3–4, 197–209.

R.J. Evans, J.R. Pulham, J. Sheehan. On the number of complete subgraphs contained in certain graphs, *J. Combin. Theory Ser. B*, 30 (1981), no. 3, 364–371.

C.F. Gauss. *Disquisitiones Arithmeticae*, Fleischer 1801.

J. Greene, D. Stanton. A character sum evaluation and Gaussian hypergeometric series, *J. Number Theory*, 23 (1986), no. 1, 136–148.

K. Hashimoto, L. Long, Y. Yang. Jacobsthal identity for $\mathbb{Q}(\sqrt{-2})$, *Forum Math.*, 24 (2012), no. 6, 1125–1138.

H. Hasse. Beweis des Analogons der Riemannschen Vermutung für die Artinschen und F. K. Schmidtschen Kongruenzzetafunktionen in gewissen elliptischen Fällen, *Nach. Gessel. Wiss. Göttingen Math.-Phys. Klasse*, 1 (1933), no. 42, 253–262.

R.H. Hudson, K.S. Williams. An application of a formula of Western to the evaluation of certain Jacobsthal sums, *Acta Arith.*, 41 (1982), no. 3, 261–276.

R.H. Hudson, K.S. Williams. Resolution of ambiguities in the evaluation of cubic and quartic Jacobsthal sums, *Pacific J. Math.*, 99 (1982), no. 2, 379–386.

E. Jacobsthal. *Anwendungen einer Formel aus der Theorie der quadratischen Reste*, Dissertation, Berlin, 1906.

E. Jacobsthal. Über die Darstellung der Primzahlen der Form $4n+1$ als Summe zweier Quadrate, *J. Reine Angew. Math.*, 132 (1907), 238–246.

G.A. Jones. Paley and the Paley graphs, in *Isomorphisms, Symmetry and Computations in Algebraic Graph Theory*, pp. 155–183, Springer Proc. Math. Stat., no. 305, Springer 2020.

C. Jordan. Sur les congruences du second degré, *C.R. Acad. Sci. Paris*, 62 (1866), 687–690.

V. Kiritchenko, M. Tsfasman, S. Vlăduţ, I. Zakharevich. Quadratic residue patterns, algebraic curves and a K3 surface, *Finite Fields Appl.* 101 (2025), Paper no. 102517.

A.A. Klyachko, A.N. Vassilyev. Balanced factorizations, *Amer. Math. Monthly*, 123 (2016), no. 10, 989–1000.

D.E. Knuth. Two notes on notation, *Amer. Math. Monthly*, 99 (1992), no. 5, 403–422.

D. Krachun, F. Petrov, Z.-W. Sun, M. Vsemirnov. On some determinants involving Jacobi symbols, *Finite Fields Appl.*, 64 (2020), Paper no. 101672.

V.-A. Lebesgue. Recherches sur les nombres, Part 2, *J. Math. Pures Appl.*, 3 (1838), 113–144.

D.H. Lehmer, E. Lehmer. On the cubes of Kloosterman sums, *Acta Arith.*, 6 (1960), 15–22.

F. Lemmermeyer. *Reciprocity laws. From Euler to Eisenstein*, Springer Monographs in Mathematics, Springer 2000.

F. Leprévost, F. Morain. Revêtements de courbes elliptiques à multiplication complexe par des courbes hyperelliptiques et sommes de caractères, *J. Number Theory*, 64 (1997), no. 2, 165–182.

B. Mazur. Rational points of abelian varieties with values in towers of number fields, *Invent. Math.*, 18 (1972), 183–266.

L.J. Mordell. On Lehmer's congruence associated with cubes of Kloosterman's sums, *J. London Math. Soc.*, 36 (1961), 335–339.

T. Nagell. *Introduction to Number Theory*, Second edition, Chelsea Publishing Co. 1964.

K. Ono. Values of Gaussian hypergeometric series, *Trans. Amer. Math. Soc.*, 350 (1998), no. 3, 1205–1223.

R.E.A.C. Paley. On orthogonal matrices, *J. Math. Phys.*, 12 (1933), 311–320.

J.C. Parnami, M.K. Agrawal, A.R. Rajwade. Some identities involving character sums and their applications, *J. Indian Math. Soc. (N.S.)*, 54 (1989), no. 1–4, 125–132.

D. Poulakis. Évaluation d'une somme cubique de caractères, *J. Number Theory*, 27 (1987), no. 1, 41–45.

A.R. Rajwade. The Diophantine equation $y^2 = x(x^2 + 21Dx + 112D^2)$ and the conjectures of Birch and Swinnerton-Dyer, *J. Austral. Math. Soc. Ser. A*, 24 (1977), no. 3, 286–295.

A.R. Rajwade. On a conjecture of Williams, *Bull. Soc. Math. Belg. Sér. B*, 36 (1984), no. 1, 1–4.

R.A. Rankin. Generalized Jacobsthal sums and sums of squares, *Acta Arith.*, 49 (1987), no. 1, 5–14.

P. Roquette. *The Riemann hypothesis in characteristic p in historical perspective*, Lecture Notes in Math., no. 2222, History of Mathematics subseries, Springer 2018.

S. Selberg. Ernst Jacobsthal, *Norske Vid. Selsk. Forh. (Trondheim)* 38 (1965), 70–73; in English: http://www.numbertheory.org/obituaries/OTHERS/jacobsthal_eng.html.

R. Siegmund-Schultze. *Mathematicians fleeing from Nazi Germany. Individual Fates and Global Impact*, Princeton University Press 2009.

Z.-H. Sun. On the number of incongruent residues of $x^4 + ax^2 + bx$ modulo p, *J. Number Theory*, 119 (2006), no. 2, 210–241.

A. Thomason. On finite Ramsey numbers, *European J. Combin.*, 3 (1982), no. 3, 263–273.

L. von Schrutka. Ein Beweis für die Zerlegbarkeit der Primzahlen von der Form $6n + 1$ in ein einfaches und ein dreifaches Quadrat, *J. Reine Angew. Math.*, 140 (1911), 252–265.

A. Weil. *Sur les courbes algébriques et les variétés qui s'en déduisent*, Hermann 1948.

A.L. Whiteman. Theorems analogous to Jacobstahl's theorem, *Duke Math. J.*, 16 (1949), 619–626.

A.L. Whiteman. Cyclotomy and Jacobsthal sums, *Amer. J. Math.*, 74 (1952), 89–99.

A.L. Whiteman. A theorem of Brewer on character sums, *Duke Math. J.*, 30 (1963), 545–552.

K.S. Williams. Finite transformation formulae involving the Legendre symbol, *Pacific J. Math.*, 34 (1970), 559–568.

K.S. Williams. Note on Brewer's character sum, *Proc. Amer. Math. Soc.*, 71 (1978), no. 1, 153–154.

K.S. Williams. Evaluation of character sums connected with elliptic curves, *Proc. Amer. Math. Soc.*, 73 (1979), no. 3, 291–299.

K.S. Williams. An arithmetic proof of a theorem of Chan, Long, and Yang, *Amer. Math. Monthly*, 130 (2023), no. 8, 760–763.

D. Zagier. A one-sentence proof that every prime $p \equiv 1 \pmod 4$ is a sum of two squares, *Amer. Math. Monthly*, 97 (1990), no. 2, 144.

Index